**Climate Change is Natural
NOT Man Made**

Sunspots = Warming

by

Leon A. Luyckx

August 2021

Published by
Leon A. Luyckx

>

Table of Contents.

PROLOGUE.

Man doesn't Make Climate.
Climate Makes Man.

	°C	Trend
• Early Hominids Climbed Down from the Trees in Africa	- 7	cooling
• Early Holocene Energizes Man: The Pyramids	+ 13	warming
• Sumerians, Minoans, Egyptians	+ 3	warming
• The Romans	+ 2.5	warming
• The Muslims: Greening of the Arabian Desert	- 2.5	cooling
• The Carolingians and the Vikings	+ 1.0	warming
• Universities, Castles, Cathedrals and Crusades	+ 2.5	warming
• Genghis Khan and his Mongols (1206 –1220)	500 years of	warming
• The 1348 Bubonic Plague (a bacteria)	- 2	cooling
• The Last 200 years: Explosion of Progress	+ 2	warming

Introduction.

Over 90% of the people of this world strongly believe that our warming climate is due to human action. The following 3 facts are undeniable:
1. Our climate is warming up.
2. Burning fossil fuels since 1850 increased air CO_2 from 280 to 420ppm or 0.042%.
3. Keeping millions of cows increases CH_4 in the air to some 100ppb or 0.00001%.

What is not factual is to link CO_2 + CH_4 to global warming. That is a theory launched in the 1960s that has never been proven. It has never been publicly challenged either. No alternative cause to the warming has ever been seriously considered and certainly not by the Intergovernment Panel for Climate Change, the IPCC. From hypothesis to thesis to belief to dogma, the greenhouse gas effect of CO_2 has become, across the world, the only possible cause of our warming. No possible alternative!

During the eighties, I watched, in dismay, the ascent of guilt inflicted on mankind while I was making a living with inventions in steelmaking. This wrongful indictment drove me, upon retirement in 2004, to embrace the climate challenge as single, post career assignment. My education as geological engineer helped considerably. In Part ONE, I show that, long before the industrial revolution, our climate was cyclical for at least 3500 years. In Part TWO, I closely connect these cycles with a fast growing science showing the sun's magnetic radiation cycles as the direct cause of our warming. Thus, man does not make climate. The sun does!

Decoupling Man from Climate.
Abandoning useless CO_2 and CH_4 abatement projects would redirect an enormous amount of human energy and resources. It is not just the trillions of dollars and euros saved. It is a state of mind: We are not guilty for the climate. We are guilty of the environment, of mishandling south to north migrations and driving animals to extinction.

The Environment.
Now that we see that climate is not man-made, it is a crime to bunch it up with the environment. I am a devoted environmentalist as evident from Part SIX. There is no question that humans are a major threat to the planet's well being, its natural habitat, its wildlife. That is where budgets should be redirected from unproductive CO_2 abatement.

The Reckoning.
The sun's magnetic energy radiation has diminished so much over the last decade that, despite a delay caused by arctic ice melting, a substantial cooling is coming. While CO_2 keeps rising, the cooling of the 2020-2030 decade will force a reckoning. The IPCC will be very creative to keep the CO_2 fiction alive despite factual evidence of divergence:
CO_2 rising and temperatures dropping.

PART ONE

CLIMATE IS CYCLICAL

Climate Change is NOT NEW,
Not Unique to our Time!

In fact, our climate on earth changes all the time not just since the industrial revolution.

Depending on the approach, climate can be wet or dry, calm or breezy, hot or cold. This book only looks at hot or cold.

Climate Cycles are of two Magnitudes:

I. The Maxi-Cycles: Period: 100,000 years. Temperatures -8° to +5°C Glacial Ages. Most of you, readers, know something about the major glaciations which have preceded our generally warm period, the Holocene of the last 12000 years. We will return to these maxi-cycles at the end of Part Two and Part Seven.

II. The Mini-Cycles: Period: 1000 years. Temperatures: -2 to +3°C Studying the climate since 2004, I became increasingly aware of history being intimately interwoven with climate. By history, I mean the last 2000 years, pushing it further but carefully, to 3000 years. Beyond that, it becomes prehistory with insufficient human information.

Over 10 years ago, the 1000year cycle became clear to me independently of any such information in the literature. It was a full chapter with figures of my first 2014 book: *Warming? Yes! Man Made? No!*[1] Only later did I find out that Dr John A. Eddy had suggested it in 1976. The 1000year mini-cycle is now called the Jack Eddy cycle.[2]

The Last 3 1000 year mini-cycles.

1.   The first 1000year mini-cycle: the Minoan High and Intermediate: 1250 to 250 BC

a) The climate maximum 1250 to 1000 BC.

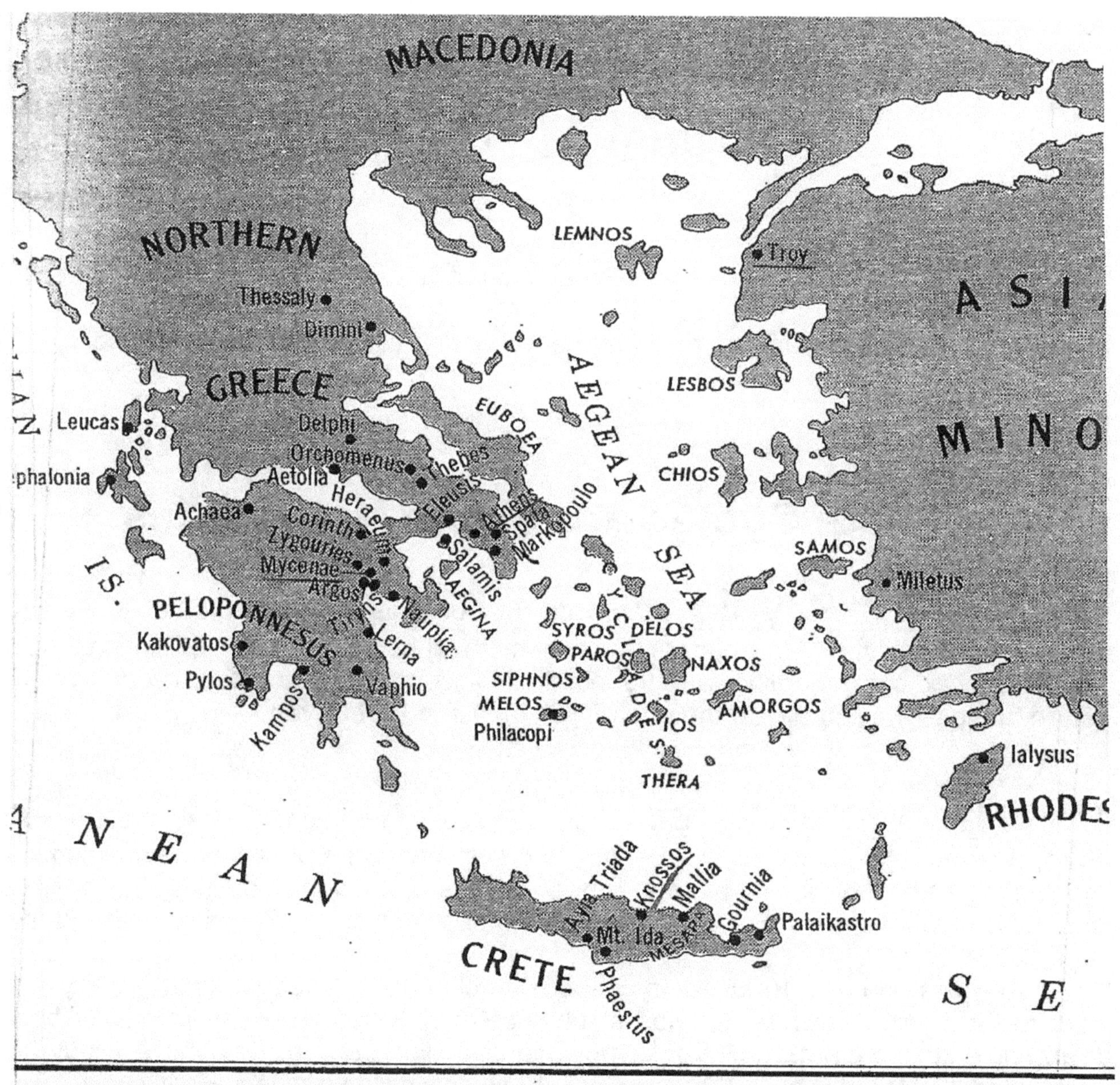

Fig 1. The above map encompasses the entire Aegean Sea civilization area over 2000 years. Please locate Knossos in Crete, Mycenae of the Lion's gate in Peloponnesus, Troy of Achilles in Northern Turkey and finally Athens in Attica. . From 1972 Encyclopedia Britannica.[3]

To permit such culture as the Aegean Sea Civilization to flourish in Crete first, advocates for a considerable warming. It was King Minos of Knossos, Crete, who unknowingly prompted historians to name it Minoan. The timing is a bit sooner than 1000 Years BC, around 1200 to 1150 BC. However, the further we go back in time, the less precise the dating becomes.

Kings David and Solomon in Israel. 1000 BC. Confirming this optimum climatic timing, we witness a simultaneous peak of glory for the people of Israel.

Pharaoh Ramses II of ancient Egypt, 1250 BC dovetails also well with the optimal climate of the time.

These first three examples of a hot climate are obviously less convincing and more blurry to date than more recent features. They are however first examples of human history influenced by climate. Warming means excess food generation allowing people to develop and expand metallurgy -the Bronze Age-, construction, art and ceramics, weaponry and other goods.

b) The Intermediate period after the Minoan optimum.

Is there evidence of a cooling between 1000 and 0 BC? I cannot find any evidence of cooling as can be shown by the following history.
Starting from the Minoans of Knossos in Crete, the next highlight are the Lions of Mycenae with Agamemnon's treasury of gold. The third Aegean star is the epic Iliadic story of Achilles and Troy. A continuous dotted line of history can be traced leading to Athens by 750 BC.

Athens: Paradoxically, the brilliant Greek civilization with its creation of democracy , philosophy, geometry and other sciences fits right in the middle of that period. The brilliance of Athens (650 – 250 BC) is not of the empire type. It is a giant leap in education and sharing which, way ahead of its time, creates the very concept of democracy. It does not require warming which is needed to feed giant armies. But there is no sign of cooling either. Attacked by Xerxes of Assyria which was an empire in expansion, the Greeks won with intelligence against their fleet in the bay of Salamis with Greek fires and sun reflecting parabolic mirrors.

Later, around 330 BC, Alexander of Macedonia does create an empire, beating and killing Darius of Assyria, going all the way to the Indus river, past Afghanistan and into Pakistan. The name Kandahar goes back to Alexander the Great. That does not fit at all with a cooling climate. So in short, no evidence of cooling from 700 to 300 BC. Rather the opposite without major warming. From now on, the 1000 year mini-cycles become very clear all the way to our present warming in 2020 AD.

* * * * *

<u>**2. The Second 1000year Mini-Cycle: The Roman Empire and Dark Age.**
250BC- 750AD</u>

<u>a) **First Half**: The Warm Roman Empire: 250 BC to 250 AD.</u>

The early wake-up call for the young Roman State came from a fierce competitor across the Mediterranean Sea: Carthage, today's Tunisia. That conflict is called the Punic wars and lasted from about 250 to 200 BC. For Hannibal and his elephants to cross the south western Alps, albeit not over a high mountain pass, the climate had to be about neutral but not cold as we developed in the third cycle. The Romans had a hard time stopping Hannibal and his Carthaginians but were finally victorious just North of Rome. Beginning with the end of the Punic wars, around 200 BC, the Roman domination of the Mediterranean basin surged at an accelerated pace.

Julius Caius Caesar was born about 100 BC [4]. Now the climate has warmed considerably and Rome was less than 100 years from becoming an empire. It is very symptomatic of a warming trend that Julius Caesar went North to Gaul and Britannia. First, he had to cross the Alps through ways which were impassable 200 years earlier. His scouts found two routes for his legions. The shorter one through the Valle d'Aosta and over the Great St Bernard Pass down to Martigny and St Maurice. The longer through the Julian Alps and Engadine. Caesar's legions moved through future France with ease because harvests were plentiful. Summers were hotter than today and winters milder. The Gaul people, the Celts, were prospering but divided, thus easy to conquer. After the original battles with Vercingetorix and others, most of them settled with the Romans. An exception was the North: "Gallorum omnium, fortissimi sunt Belgae" De Bello Gallico, 59 BC [4]. However, even Boduognat and his "Nervii" (today's Hainaut in Belgium) submitted to Caesar. He then crossed the Channel and moved North through Britannia all the way to the Scottish border. He gave up on the "pics" today's Scotts. Much later, Emperor Hadrian built a wall where land was the narrowest South of Scotland.

Caesar's legions found vineyards as high as today's Manchester and, of course, plenty of food all along. That was the key to the success of the entire northern campaign.
After Caesar and many other campaigns around the Mediterranean, the Empire peaked around 100 AD and slowly declined with the onset of the next cooling after 250 AD.

<u>b) **The Second Half.** The Dark Age. 250 – 750 AD.</u>

The drop in temperature between 250 – 400 AD does not appear from all records to have been abrupt. From the climate prospective the following question is essential.

Did the Rhine River freeze over in the dead of winter 406 or 405 AD?

No chronicler of the time ever mentioned that possibility but none of them were on site, not even close! One key historian, Jerome, wrote from Bethlehem, Judea.   In addition, after more than 500 years of warm climate no historian had any idea of what a cold spell could mean to large rivers: a freeze over. It was clear, however, to all these chroniclers that a stone bridge crossed the Rhine in today's Mainz (Mogundiacum). They all assumed, of course, that the invading barbarians crossed over that bridge by the tens of thousands. This is very unlikely. That bridge was guarded by a maximum concentration of Frank "Federati" soldiers. In addition, all chroniclers agree the invasion happened in the dead of winter. If it was over a bridge it would have been summer. On the other hand, once a large river freezes over, it does become solid for tens of miles, opening the choice for massive crossing on foot with chariots, baggage and farm animals at a stealth, unguarded location. Any chance observer on the other side could be easily silenced. So, the following quotation, although controversial, summarizes what most likely happened. "For, on the bitterly cold night of December 31, 406, there was apparently no Roman army on guard where a host of Vandals, Alan, Suevi and Burgundians with their families and possessions crossed the frozen Rhine and headed southwest through Gaul. This time, the Roman frontiers had been breached by barbarians who meant to stay."[5]

In short it is almost certain that the Rhine froze over in the early 400s. So, most likely did the Thames and other rivers. The barbarians crossed the frozen Rhine possibly over several winters to explain the size of the invasion.

The most characteristic modus vivendi of the Dark Ages.

In Western Europe the Dark Ages were centered on surviving the cold and the lack of food. The saving grace were newly founded Monasteries started by Irish missionaries. They built large, structurally sound establishments with plenty of storage room for grain, fruit and other winter staples. They also housed cattle, pork and poultry. That in turn, prompted villagers to live around them at walking distance to survive consecutive years without harvest. Christian monks were thus the critical linchpin to dark age survival.

The Muslims: The one example of inverse climate effect.
It is also during the deepest part of the dark age around 600 AD that the Muslim civilization was born. Why?
I posit that while the northern temperate regions suffered deep cold, the Arabian desert flourished with more humidity and coolness than usual. Medina and Mecca on the Red Sea could sustain a fast growing population with almond trees, fruit, grains and vegetables. Less tension from food shortage led to more cooperation between Arab tribes, at least temporarily. Long distance camel travel became easier and less deadly.
For Caesar, the expansion North was climate driven. For the Muslims, the Northern move was driven by the knowledge of old treasures in Damascus, Jerusalem, Cairo and Constantinople. The latter one, though, resisted successfully for 800 years until 1453. The Muslims quickly extended their conquest along the North African coast, today's

Egypt, Libya, Tunisia, Algeria and Morocco. That territory was also greener and more food productive than today. By 700, they had conquered the "Andalus" and established a Caliphate in Cordoba, Spain, replacing the old Visigoth kingdom. Their limit was the Pyrenees, a welcome natural barrier for the Franks.

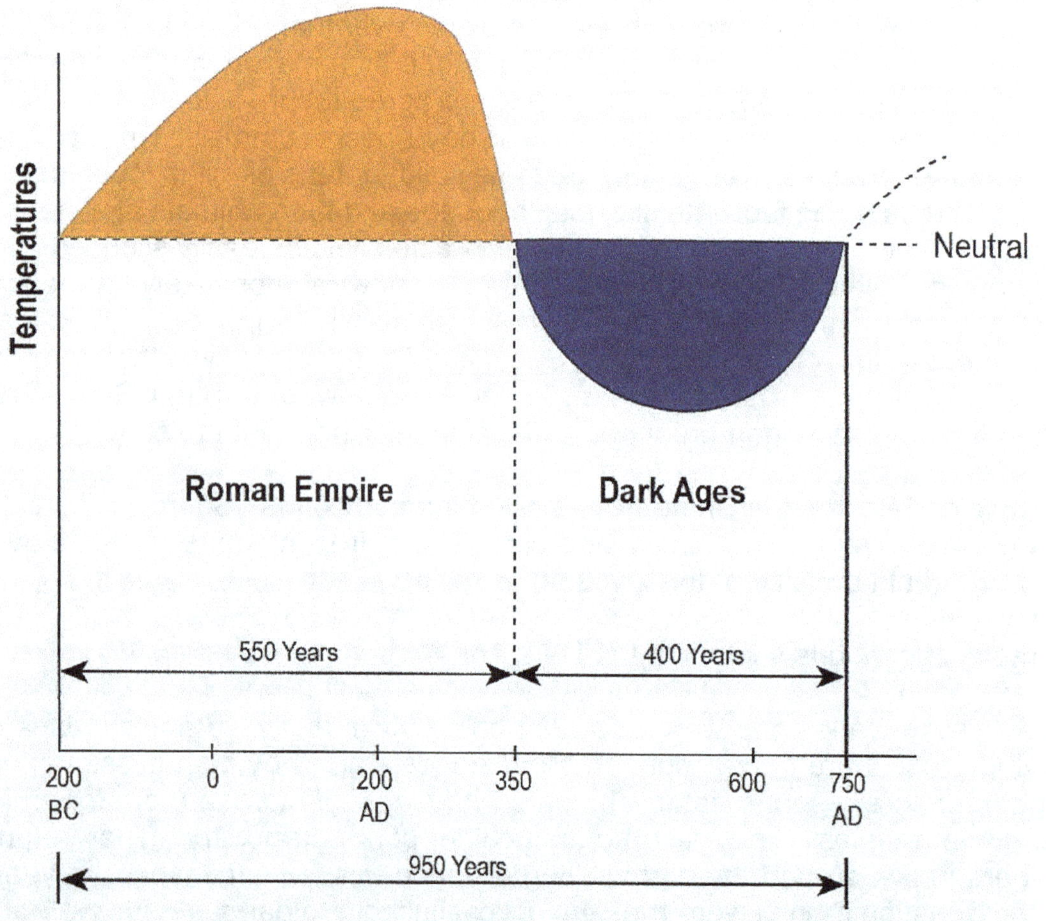

Fig.2 Simplified and schematic temperature-time diagram covering the second 1000year mini-cycle. In my 2014 book,[1] from which this diagram is taken, I still thought the cycle was 950 years, pretty close.

And so, from 250 BC and the first Punic wars, we have reviewed the second mini-cycle, ending with the end of the dark age in 750 AD. This second mini-cycle is in fact the first one with a clearly cool second half, the Dark Age. The first, Minoan mini-cycle had no appreciable cooling from 750 BC to 250 BC.

* * * * *

<u>**3. The third 1000year mini-cycle: The Middle Ages & the Little Ice Ages, 750-1750 AD.**</u>

a) <u>**First Half**</u>: <u>The Medieval Warm Age. 750 – 1250 AD</u>.

The more we advance, the more definition appears in chronicles, sagas, centers for learning, monasteries and the first universities. It is still all handwritten, recopied and illustrated. The warming of the air, the return of good harvests, the retreat of ice in the North Atlantic generated unprecedented activities in Europe, Southern Mexico and Mongolia at the end.

The first big wake-up call: <u>The Battle of Poitiers – Tours 736AD.</u>

The onset of the warm period to come made the Pyrenees finally passable for a large army of "Saracens". The grandfather of Charlemagne, Charles Martel (the Hammer) extinguished the Muslim's desire of conquest in northern Europe. The great battle and pursuit back to the Pyrénées took several weeks. The destruction of the Saracen army, the capture of their gold and treasury was almost total. Only a few hundreds of Saracens crossed the Pyrenées back to Spain. Like with the Punic wars, it was again the early warming of the climate transition that had driven the Muslims north but this time they hit a wall, the Franks, an overwhelming force.

<u>The Carolingian Civilization, 736 – 814AD</u>. Brief but spectacular.

Europe was barely pulling out of the long Dark Age, shaking up centuries of hunger and hardship. Charlemagne, son of Pepin le Bref was born in Herstal on the river Meuse, just north of Liège. He was a gifted leader , warrior and educator. With a well equipped army of second and third generation soldiers started by his grandfather the Hammer, Carolus Magnus swept through Europe, baptizing Saxons by the hundreds in the Rhine. He reached Poland, Hungary and Austria. On Christmas day of the year 800, he was sacred "Emperor of the Roman Empire" by Pope Leo III in Rome. That was it! After his death in 814, his son Louis the Pious, let everything crumble to division and decay. What was left was a strong message of education for simple people.

<u>The Vikings: 760 to ~ 1200 AD</u>.   Master Warriors of the North Atlantic Ocean.

In his old age, from his castle along the river, Charlemagne watched the first "drakkars" (long boats) full of Viking warriors rowing up the river invading the Lowlands.

Unlike the one exceptional lineage of Pépin and Charles resulting in a brief civilization, the Vikings included many coastal communities of Norsks (Norwegians today) and Dansks (Danes today). **Their influence, peaking from 800 to 900 in the lowlands kept moving further north with the warming climate of the Middle Ages.**
The Dansks came first in Normandy and the Lowlands, 763 to 892. The Norsks attacked Ireland and Britain starting a bit later and moved North to the Orkney, Hebrides and

Faroes Islands ending up in Iceland. The ice continuing to recede North, the most adventurous Eric the Red and his son Leif Ericson settled Greenland and pushed all the way to Newfoundland at Anse aux Meadows in about 1000 AD.

In the recent climate literature [6], a newcomer has appeared: the Oort minimum. It is reported to be a deep cooling between 900 and 1000 AD. From the previous story underpinned by warming this Oort episode would appear at odds. The reality may be closer to what will later be called a Dalton minimum in Part Two. This is a minor 30 to 40 year cooling of about 1° C. So, the Oort minimum was probably a Dalton "mini-cooling" in the mid-900s.

The Mexican Rebirth of the Mayan Civilization. 900-1100 AD.[7]

Old Mayan monuments are found south of Mexico in Guatemala, dating back to 292AD. Then a long hiatus occurs from 300 to 900AD. Finally during our early Middle Ages, the Mayan rebirth occurs much further north in Mexico. The architecture and art keeps revealing new underground extensions.

The Grand Maximum of the Middle Ages, 1000-1250AD.

Everyone of these medieval stories is strongly climate driven. It is striking how much activity occurred simultaneously, each one requiring an enormous amount of human effort, team work and sweat. The key to all of that was **plenty of food and warm winters.** The climate favored a massive growth of population needed for all these achievements. In architecture for example the building of castles and cathedrals required an astonishing array of new trades: masons, stone cutters, stone carvers, metal smiths, stain glass makers, tool makers and carpenters leading to the future corporations. In addition to this construction of castles and cathedrals, there was a simultaneous enterprise of travel to Jerusalem started in 1096 by Pope Urban II: The Crusades. This extracted an enormous amount of people away from local work. On the other hand, it educated everyone to the geography of Europe and the Middle East. On average far less than 50% came back home. The very first crusade by Pierre l'Hermite and his poor crusaders lost 95% of its people in today's Turkey.

At the same time philosophy, rhetoric, dialectic and theology were concentrated in new centers of learning, the first universities. The first one was in Paris, the future Sorbonne. The second was in Oxford, UK.

To summarize this medieval miracle of human accomplishments, here are the Highlights.

1/ 736 – 814: Carolingians, Charles Martel and Charlemagne.
2/ 763 – 1050: The Vikings follow the retreat of the ice in the North Atlantic.
3/ 1096 – 1250: Crusades made possible by the warming.
4/ 1150 – 1300: Gothic Cathedrals, enormous undertaking.
5/ 1150 – 1300: Feudal Castles, late reaction to the Vikings.
6/ 1100 - …   : First Universities, Paris, Oxford, Bologna, etc

<u>The Last Medieval Event: The Mongols and Genghis Khan. 1200 – 1260.</u>

Please note the dates: this is emphatically climate-driven. Mongolia is such a desolate high plateau that, most of the time, it can only support a small nomadic population, lodging in movable "yurts" and surviving on domesticated animals. **This is why it took the whole 450 years of Medieval warming** to build a population of warriors, thousands of small horses, movable food and lodging, light weaponry and, finally, a strong, unscrupulous leader to conquer central Asia. Named Khan by his subjects in 1206, Genghis Khan first ravaged sleepy northern China. Then he turned west and butchered Samarkand and Bukhara in today's Uzbekistan. Those short horsemen went all the way to Poland, Ukraine, Iran and Bagdad in Iraq.
The abrupt climate change starting around 1250 put an early end to this "First Mongol Empire". Born in 1336 near Samarkand, Timor Leng is said to have built the second Empire but despite resounding victories, he did not carry the vastness of his great grandfather's conquests.

Indeed, the extensive warming of five medieval centuries triggered at least four new civilizations over several continents. It is only our ignorance of Chinese history that prevents us from finding similar medieval revivals in that vast multi-millennial empire.

b) **The second half** : The Little Ice Ages. 1250 – 1750 AD.

After the warm Middle Age which saw an amazing jump in multi directional activities, it all came crashing down within a few decades because of sudden cooling.

<u>The wake-up call: The Mesa Verde Abandoned. 1250 – 1280.</u>

Dated by tree ring proxies, it is now well documented that the extensive masonry construction by the local Indians peaked around 1200 – 1250 AD. Around 1260 – 1280, this feverish activity was abandoned abruptly without any destruction. So, without trace of battle or enemy attack, the population left probably to greener pastures. A sudden change in climate must have been the primary driver of this event.

<u>The Steep Temperature Drop in Western Europe. 1300 – 1348.</u>

Old chronicles, particularly from Brittany, report consecutively harsh winters. These were followed by wet and cold springs, with fruit flower frosts and summer grain harvest failures. This resulted in hunger, sickness and death. The contrast with the plenty throughout the 1000 to 1200 AD period could not have been sharper. The hope and prayers were that it would end soon but it did not. The climate inversion was in to stay. To survive those harsh winters in the primitive, drafty, uninsulated wooden homes of the time, a new approach was taken everywhere: keep the farm animals in the house on the ground floor and sleep upstairs, benefitting from the warmth of those natural heaters. Rats, mice and other critters were soon part of the household. A more perfect setting for the quick spreading of a pandemic can hardly been imagined.

The Black Death.   1347 – 1349.

Adding insult to injury, a furious pandemic of bubonic plague invaded Europe at lightning speed. It started in 1347 from a south Italian port, probably Naples, originating from the Near East (Le Levant). The pandemic jumped by leaps and bonds across the European continent.   We know now that this was not a virus. It was a bacteria transmitted mostly by lice on rats and other small mammals. It spared very few places. Some small towns were unexpectedly intact while most others lost heavily. The well accepted estimate is that at least one third of Europe's population of the time died from the plague. The loss was added to unknown mortality surges since about 1300.

The End of the Vikings in Greenland. 1300 – 1400.

In Collapse[8], Jared Diamond describes at length the heroic survival of the Norsk Vikings settled in southwest Greenland after 1300. Despite being accustomed to hardship, the drastic cooling removed their ability to survive. By 1410, the last Danish ship with supplies had reached Greenland. The next Danish sailing was in 1728. It is known from that reconnection that the population of Greenland completely died out by 1450.
If western Europe was already so deeply affected by the cooling, it must have been atrocious in the Greenland settlements. As Iceland is concerned, the combination of the Gulfstream and a volcanic substrate permitted the survival of about half the medieval population.

From a climate view point, **this was the first of 3 mid-millennium Little Ice Ages**. In Part II, we will name this first temperature dip, the "Wolf", 1300 to 1365.

Limited Warming between Two Little Ice Ages. 1365 - 1450.

In the chronicles, the stories of harsh winters and failed harvest disappeared. Europe is rebounding in population after the big losses of the early 14th century. It is towards the end in 1440 that Guttenberg invents the printing press, revolutionizing written information. Worth mentioning also, because of the coincidence in dates, is the meteoric rise of the Dukes of Burgundy,[9] Philippe le Hardy (the Bold), Jean sans Peur (without fear), Philippe le Bon (the Good) and, tragically, Charles le Téméraire (the Foolhardy) killed at the battle of Nancy and having fathered only one daughter. The entire house of Burgundy was up for grabs.   Between 1420 and 1460 under Philip the Good surnamed "Grand Duc d'Occident", the entire future Belgium and Holland attained an apogee of wealth, fame, culture and self-governance.

A Second, Barely Perceptible Little Ice Age. 1450 – 1540.

As compared to the first and third little ice ages, this second one is hardly perceptible. There is no clear evidence of hardship, no pandemic and no loss of population. It is Renaissance time in Italy with the fame and opulence of the Medici family. It is also corruption time at the Vatican, triggering young monk Martin Luther to pin his 95 objections at the door of the Chapel of Wittenberg in Germany. Last, but not least, Henry the Eight secedes from Rome and starts the independent church of England.   Rather than a little ice age the period looks more like a lower climate plateau.

<u>A Somewhat Warmer Episode. Still Below Neutral. 1540 – 1610.</u>

Again, from the climate view point, this 70 year period is not outstanding. It is mild. Towards the end, lowland painters depict snowy winters but no solid ice on the ground: a major distinction from the solid ice in paintings of the next period.

The Hanseatic League.

One significant sign of some warming was the fast growing sea commerce on the North and Baltic Seas.   The ports of London, Dover, Antwerp, Rotterdam, Oslo, Hamburg, Bremerhaven, Copenhagen, Stockholm, Helsinki and Danzig were the sites of intense trading activities. The conversations between seamen and port crews were so dense that they evolved a spontaneous Germanic slang which no one else could understand. The prosperity of Northern Europe greatly benefitted from that Hanseatic phenomenon.

<u>The Deepest and Originally Named Little Ice Age. 1610 - 1750 AD.</u> Before we delve into this climatic disaster, let us find out when and who first introduced the very name : Little Ice Age.   It is F. E. Matthes in 1939.

Fig 3. Picture of Francois Emile Matthes from Wikipedia

It is amazing how recent the very idea of a "mini" mid millennium cooling is. The first currently known reference to "Little Ice Age" dates back only to 1939, barely 80 years ago. For example, the 1972 Encyclopedia Britannica is totally silent on the Little Ice Age to the point of not even mentioning it in its biography of F. E. Matthes[10] who first coined the appellation in 1939. However F. E. Matthes made no mention of any connection between his Little Ice Age and the sunspot minimum. Chances are he had no idea of the 300 year old sunspot science. Nobody had made the link until 1976, see part TWO.

<u>The Wake-Up Call: The 30 year War. 1618 – 1648.</u>

Starting in 1618, a multi-state conflict engulfed the entire center of Europe. It included Sweden, Denmark, Germany, Poland, Austria and the Netherland. It soon turned ugly. Winter hardship, poor equipment and training fatally injured thousands of soldiers. The 1648 treaty of Munster, however, introduced for the first time, the modern concept of nations having rights over monarchies. It also included the first appearance of human rights.

<u>The Bottom of the Little Ice Age. 1650 – 1715.</u>

By 1650, the cold spell deepened and kept deepening all the way to 1715. Consecutive winters saw the Thames, the Rhine, the Maas, the Danube and the Rhone freeze over for months at the time. There were even reports that the Nile froze over for a while. Spring frosts annihilated many fruit harvests and eliminated wineries even in Italy and Spain. Many summers produced almost no grain harvest. It was clearly a situation similar to the early 1300. At the deepest , around 1695, glaciers of the Alps grew and advanced so much that, near Chamonix, "La mer de glace" the major Mont Blanc glacier, crushed and obliterated two small villages of the French Alps.

Fig 4. Winter Landscape by Peter van Haelen (1612-1687) selected among Belgian painters because this was produced during the 1645-1700 deepest Little Ice Age. Van Haelen's signature can barely be seen on the bottom right. Copy made about 100 years ago by my grand uncle Edmond de Hepcee. I inherited this from the Walloon branch of my family.

Greenhouses, Winter Endives, Sugar Beets.

Survival strategies for food included the invention of small, lean-to greenhouses against the south facing wall of the house. It only used a precious few flat glass panels. They grew valuable vegetables year-round. Another is the "Belgian endive". Growing in the dark on a carrot shaped root in a 6 foot deep space in the ground, this whitish winter vegetable became popular during those harsh winters. Of course, the underground space was well insulated by a wood cover and hay, and later, warmed by a stove. The carrots were regenerated by planting them in good soils during the summer. It is also the approximate time when hardy sugar beets started supplying alternate energy to failing grain harvest.

<u>Epidemic Aggravation.</u>

In 1666 and following years, the black plague reappeared but, apparently, less severe on the population than in 1348. Over 300 years, hygiene had improved considerably. The lack of food and cold winters have on their own such severe effect on life that the plague's number paled in comparison.

<u>Population Loss.</u>

Hard to imagine today, the loss is estimated to be on the order of 15 to 20 million west Europeans: 5 in France, 3 in England, 6 in Germany, 3 in the Lowlands, 2 in Italy and Spain. And we don't know about Scandinavia, Poland and other east European countries.

Across the Atlantic, the 1620 pilgrims to Plymouth only had 25 years of relatively moderate climate to adjust to the local farming before very harsh conditions set in. No wonder the   population barely survived these early New England decades.

Finally, it is ironic that Louis the Fourteenth called himself "Roi Soleil" (King Sun) while his reign almost exactly covered the little ice age with heavy population losses due to hunger and disease. It is most likely that this climate disaster prepared the future French revolution a few decades later.

This marks the end of the 500 years long Little Ice Ages, 1250 to 1750. It is also the end of the third 1000 year cycle, 750 to 1750. The diagram of the following page summarizes this third cycle and shows the beginning of the fourth 1000 year cycle: Our current Little Warm Age.

* * * * *

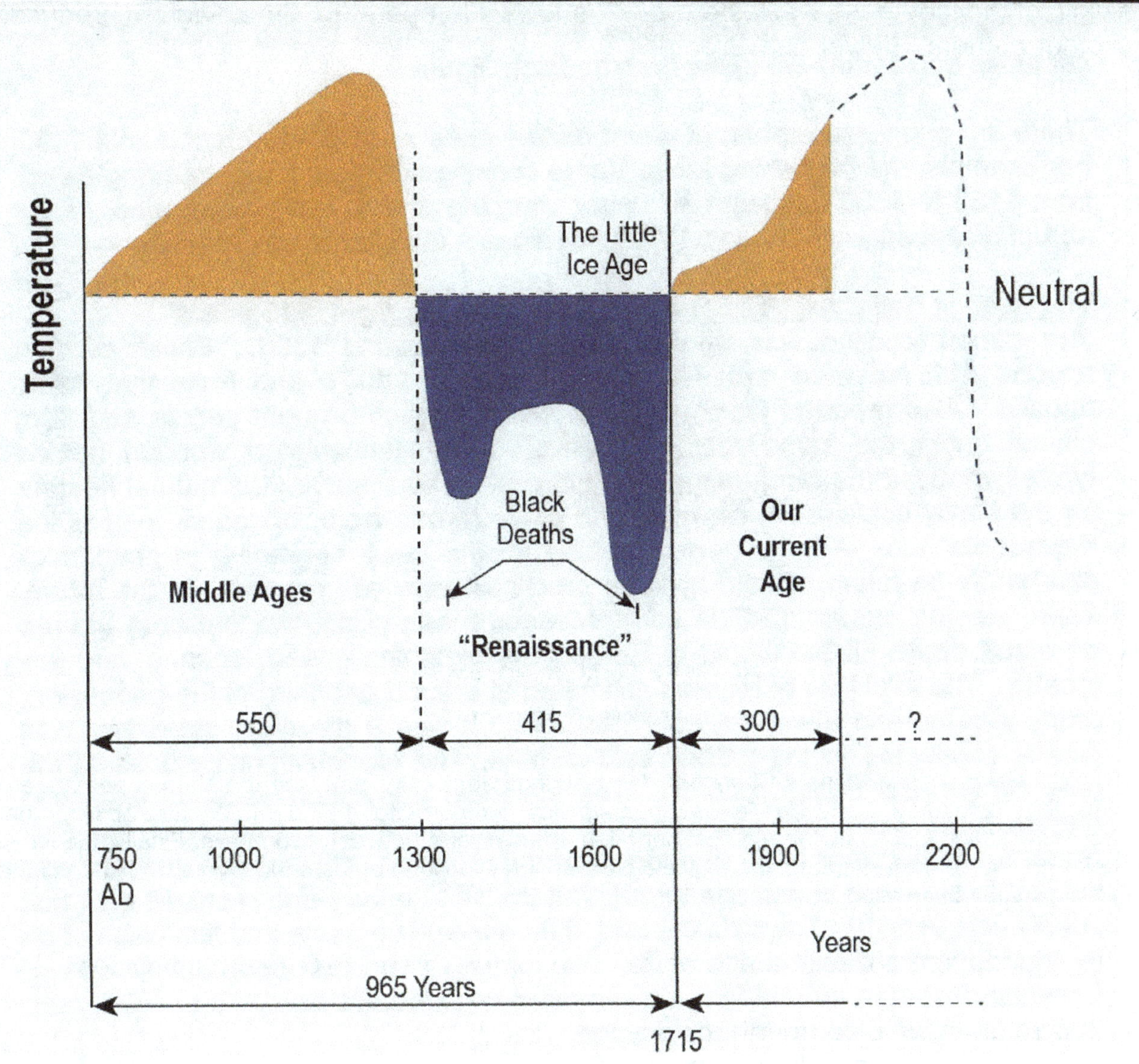

Fig 5 Schematic diagram of third and fourth 1000 year mini-cycles. Instead of 1300, we now take 1250 for the warm to cold transition. Remember the Mesa Verde. Instead of 1715 we now prefer 1750 which is the real time of perceptible warming as we will ascertain at Glacier Bay National Park in Alaska in the following pages.

The fourth mini cycle shows warming since 1750 continuing all the way to 2250 before the next probably abrupt cooling.

Now that we reviewed three 1000 years- year cycles back to back from 1250 BC to 1750 AD, let us learn what other authors have been able to reconstruct in term of temperature proxies. In his 2019 book entitled "The Solar Magnetic Cause of Climate Changes and Origin of the Ice Ages", Dr Easterbrook[11] shows a temperature time diagram produced by 2 other authors, Stuiver and Groote.

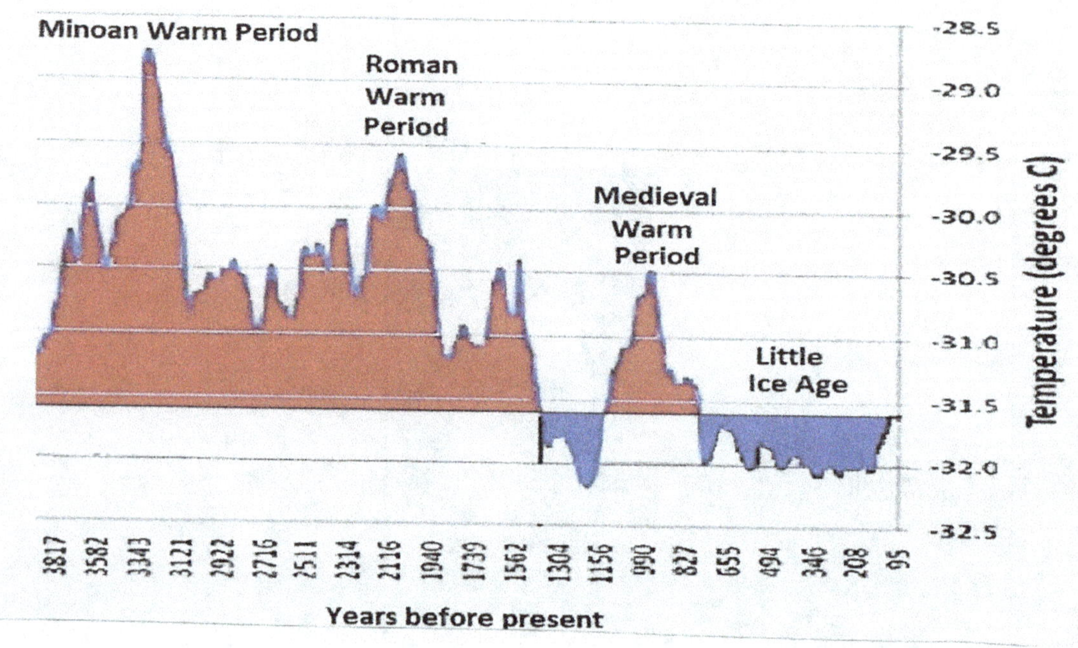

Fig 6 . Late Holocene Climate Changes. As described earlier, we now see the Minoan peak at around +3°C, the Roman at 2°C and the Medieval peak too low compared to other data. We also observe that the Greek civilization did not occur in a cool period, only less warm between two peaks. Finally, the little blue Dark Age appears too short and the multiple blue ice ages too long at the end. Overall, a fair confirmation of the climate observations.

This figure summarizes a temperature trace for the three 1000 year mini-cycles.

Let us return now to our fourth mini-cycle:
This is our mini-cycle. We are about 270 years into it, just past the middle. We can now safely predict another 200 years of warming after an impending 30 years of cooling. This short cooling parallels the Oort cooling in the early middle ages as discussed previously.

a) The first Half of the mini-cycle. Our Little Warm Age. 1750 – 2250.

After the deepest of those 3 mid-millennium little ice ages, a slow recovery occurred after 1750. However, most other climatologists ignore any kind of warming before 1850. This view obviously dovetails with the greenhouse gas theory which claims that our warming started with the industrial revolution and the rise of CO_2 in the air.

24

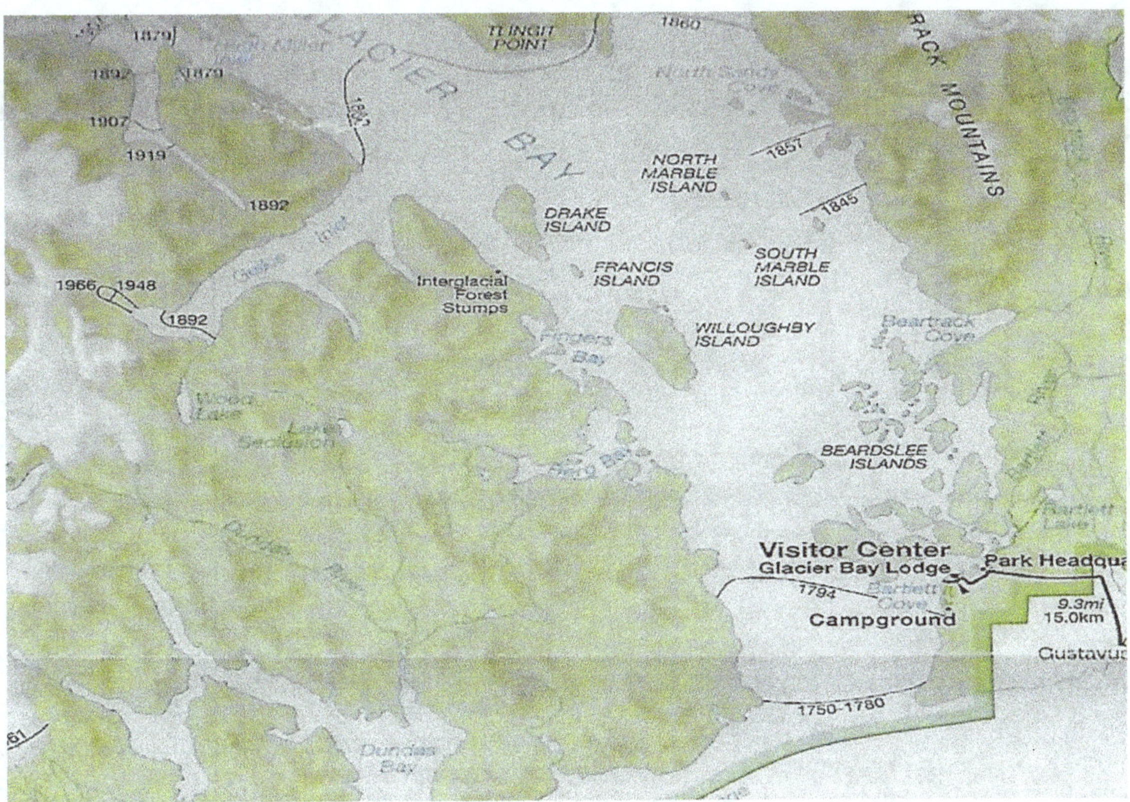

Fig 7. The retreat of the main glacier in Glacier Bay National Park, Alaska started at or before the first line marked 1750-1780 at the entrance of the Bay. (bottom right of the map). The next limit line observed by Captain Vancouver in 1794 touches the shore at the Glacier Bay campground. Then, miles in, come the 1845, 1857 lines and the big 1860 line. It is thus obvious that glacier's retreat started well over 100 years before the industrial revolution. Today the glacier has retreated all the way to the end of the bay, about 65 miles inland.

It is by accident, during a 2008 cruise of the Inside Passage in Alaska that a strong evidence of earlier warming fell in our lap. When sailing close to the entrance of Glacier Bay, we were given a brochure of the National Park. We were not allowed to go on shore but, in that brochure we discovered the evidence of warming, on the map hereabove.

We were so intrigued by the whole issue that, during the 2014 summer after the publication of my first book, my wife Claire and I took a trip to Glacier Bay. We spent 4 days at the Lodge and discussed the issue with the Park Scientists and Rangers.

The Park had issued a new brochure with stunning evidence that the glacier reached its maximum around 1750 and **retreated 5 miles by 1794.** Tinglit Indians had transmitted a story of the glacier suddenly advancing at the speed of a running dog.

At Glacier Bay you can witness geologic processes and change usually barely noticed in the span of a human life. Compare this diagram with the 1680 Huna Tlingit scene on the other side. There was no Glacier Bay then, only a broad valley with a glacier moving down it.

The Little Ice Age came and went quickly by geologic measures. By 1750 the glacier reached its maximum, jutting into Icy Strait. But when Capt. George Vancouver sailed here 45 years later, the glacier had melted back five miles into Glacier Bay—which it had gouged out.

When conservationist John Muir traveled here in 1879 the glacier had retreated 40 more miles up the bay since Vancouver's visit. A renowned author, Muir captured the popular imagination about Alaska, attracting tourists to Glacier Bay. Like most people today, they came by ship.

Fig 8. This self explanatory figure says it all: warming started in 1750.

To my knowledge, the Glacier Bay evidence is fairly unique around the world. The Alps do not offer such clear story. Its first glaciologist Louis Agassiz, 1807-1873, came too late to observe the warming of the 1750-1850 period. The unique history of the retreat of the glacier at Glacier Bay summarizes the whole period of our Little Warm Age since 1750. It also negates the current view of the Little Ice Age continuing to 1850.

Our Little Warm Age started in 1750, at least 100 years before the Industrial Revolution.

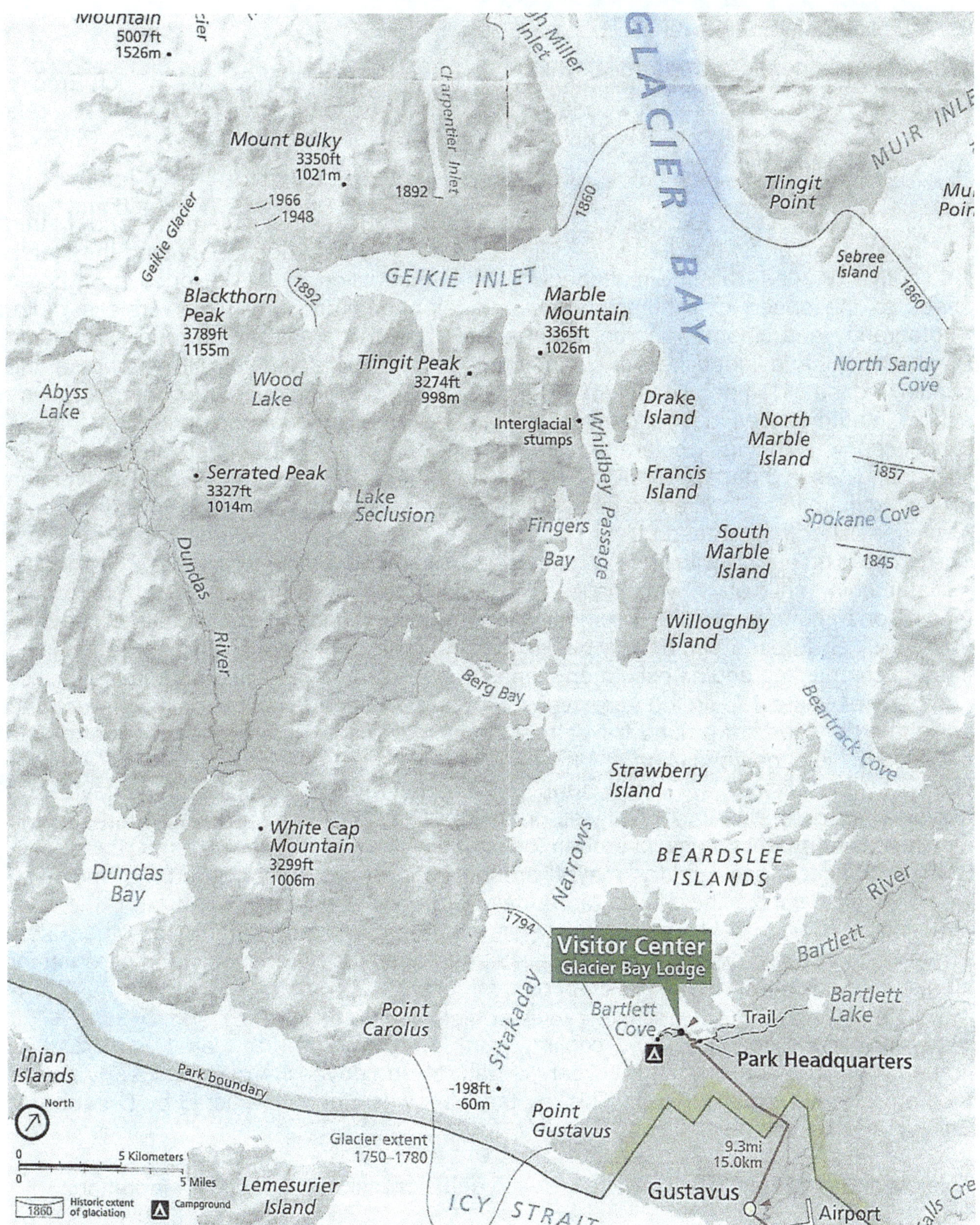

Fig 9. This new 2014 map shows at the very bottom the 1750 glacier extending 5 miles into the icy straight, the key modification to the first map.

* * * * *

<u>Modern Warm Age Highlights,   1750 – 2250.</u>

The French Revolution, the American Revolution, the Steam Engine, the French Empire (1804-1815), the first trains on wooden rails, Belgian Independence (1830), the Apogee of Classical Music: Mozart, Beethoven… Geology, Life Evolution, Chemistry, Physics, Astronomy, all flourish after 1750.

1850:   The Industrial Explosion: Steelmaking by Bessemer, Siemens and Martin skyrockets from a few hundred to a million tons per year in a decade, putting the steam locomotive on steel rails. Simultaneously, communication moves at once from horse speed to the speed of light, telegraph and telephone and by 1910, wireless radio. Photography and cinema records visual events and the gramophone records sounds. Electric power and light ushers in clean streetcars and trains and so retire the horse, the candle and the oil lamps. The Internal Combustion Engine with concrete or asphalt roads ushers in automobiles independent of rails and fast trucking of goods on long distances. Transoceanic shipping accelerates with coal fired steam, then bunker oil steam, then Diesel engines in 3 decades. Mining of industrial metals is boosted by Nobel's invention of TNT dynamite. Concrete reinforced by steel bars allows buildings to sky scrape. Airplanes usher in air transport.
The list goes on and on with nuclear, television, semi-conductors, computers, cell phones, social media, antibiotics, vaccinations, artificial intelligence, robots, soil fertilization, landing on the moon, satellite communication and observation. Somewhat later, beyond 2020, we can foresee self-driving trucks, electric battery powered vehicles, retiring the internal combustion engine except on planes & ships, better recycling of metals, plastics and general waste, the profound reinvention of office work and schooling by internet and social media; speedier ground travel in highspeed "tubing". It is China now that drives the second industrial revolution.
Driven by the climate, we move north to Canada, Alaska, Siberia, Scandinavia and maybe, the South Pole under Chinese pressure. Possibly also, a breakthrough in clean electric power production by hydrogen fusion.
Somewhat later in the century, maybe around 2050, a self-sufficient team of men and women may colonize the planet Mars via the moon. Powered by the sun, the CO_2 atmosphere of Mars will be split into oxygen to breathe and carbon for many other uses. There is, however, a distinct gap in supplies: hydrogen for water supply and nitrogen for plants. I do not know yet how this can be overcome.

Starting after the second Dalton cooling, from 2050 to 2250, the planet Earth and its human voyagers face another 200 years of natural, sun driven warming as clearly shown by the previous diagram. Fig 8. This, by the way, was clearly predicted by Dr. John A. Eddy in his 1976 paper.

This continuing warming will require creative and increasingly proactive adaptation.

* * * * *

CONCLUSION of PART ONE.

We have reviewed three one thousand years mini climate cycles and the beginning of the Fourth:

I.   The Minoan Peak of $3^\circ C$ followed by the cooler time of the Greek, Athenian civilization: 1250 to 250 BC.

II.   The Warm Roman Empire of $2.5^\circ C$ with its extension to Northern Brittany. 250 BC-250 AD. Followed by the Dark Age: 250 AD to 750 AD.

III.   The Warm Middle Ages: 750 AD to 1250 AD. Followed by 3 Little Ice Ages 1250 to 1750 AD.

IV.   The first 270 years of the fourth 1000year minicycle. 1750 to 2020 AD. Our Little Warm Age.

* * * * * * * * * * *

The next part, Part TWO, will demonstrate, with the current state of knowledge, what a growing number of climatologists believe is causing these one thousand year mini-cycles: The sun's magnetic radiation through sunspots.

These climatic mini-cycles of the last 3000 years do not support at all the recent dogma of CO_2 causing warming, which dominates today's climate science. This will be reviewed in Part THREE.

PART TWO

What Drives Climate Mini-Cycles?

THE CYCLICAL MAGNETIC RADIATION OF THE SUN

THROUGH

SUNSPOTS

What is a Sunspot?

Sunspots are darker areas, often in a group, on the bright surface of the sun as in the picture below. Called the photosphere, the sun's surface is at a constant 5,500°C. Sunspots are cooler, down a few hundred degrees as in the picture below or as much as 1500°C down in the big black spots. Sunspots are temporary, lasting from a couple of days to several months. They move from east to west with the surface of the sun in 14 to 17 days. The big black spots, such as the ones on the front cover of this book, are visible with the naked eye. This, however, can only happen when fog pales the sun to a white disk. Thanks to morning fog, at least twice in November 1952, I was surprised to observe a big black dot plus a close by shade on the center left of the sun's disk.

Fig 10. One little sunspot on the left and a group of some 5 or 6 sunspots catalogued AR 2736 were photographed on march 20, 2019. These brownish spots were small, weak and very short lived, less than 2 days. This is typical of our current period of weak solar activity.[12]   http://www.solarham.net

We are not going to see big sunspots for the next 30 years. This will be developed in point 11 of this part.

The next two pages exhibit the very best examples I found anywhere of large sunspots. I owe this to my grand aunt, Marthe L'Hoir who gave me an old 1891 astronomy book for my 15th birthday in April 1948. This book contributed to my engineering career and prepared me to understand sunspots observed 4 years later with the naked eye.

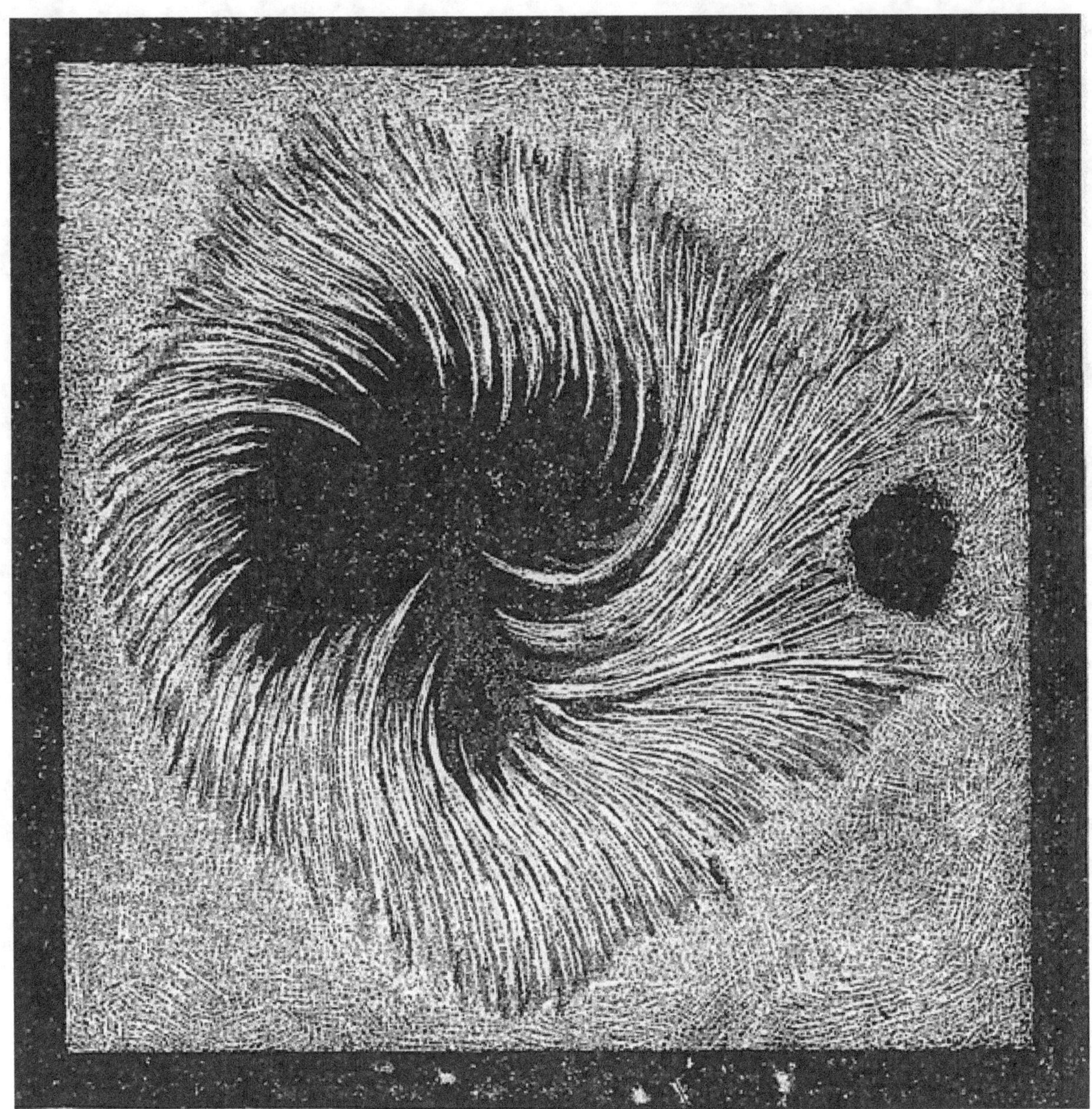

Fig 11. Before photography, the astronomers at the National Observatory of France at Meudon near Paris drew with great precision the telescoping image of sunspots. I selected this one, about the size of earth, because of very distinctive "luminous filaments" spiraling towards the center, suggesting their strong magnetic nature to be discovered in 1908 by George E. Hale. From "L'Astronomie Populaire".[13]
A 1891 book by Camille Flamarion, page 330. Enlarged twice.

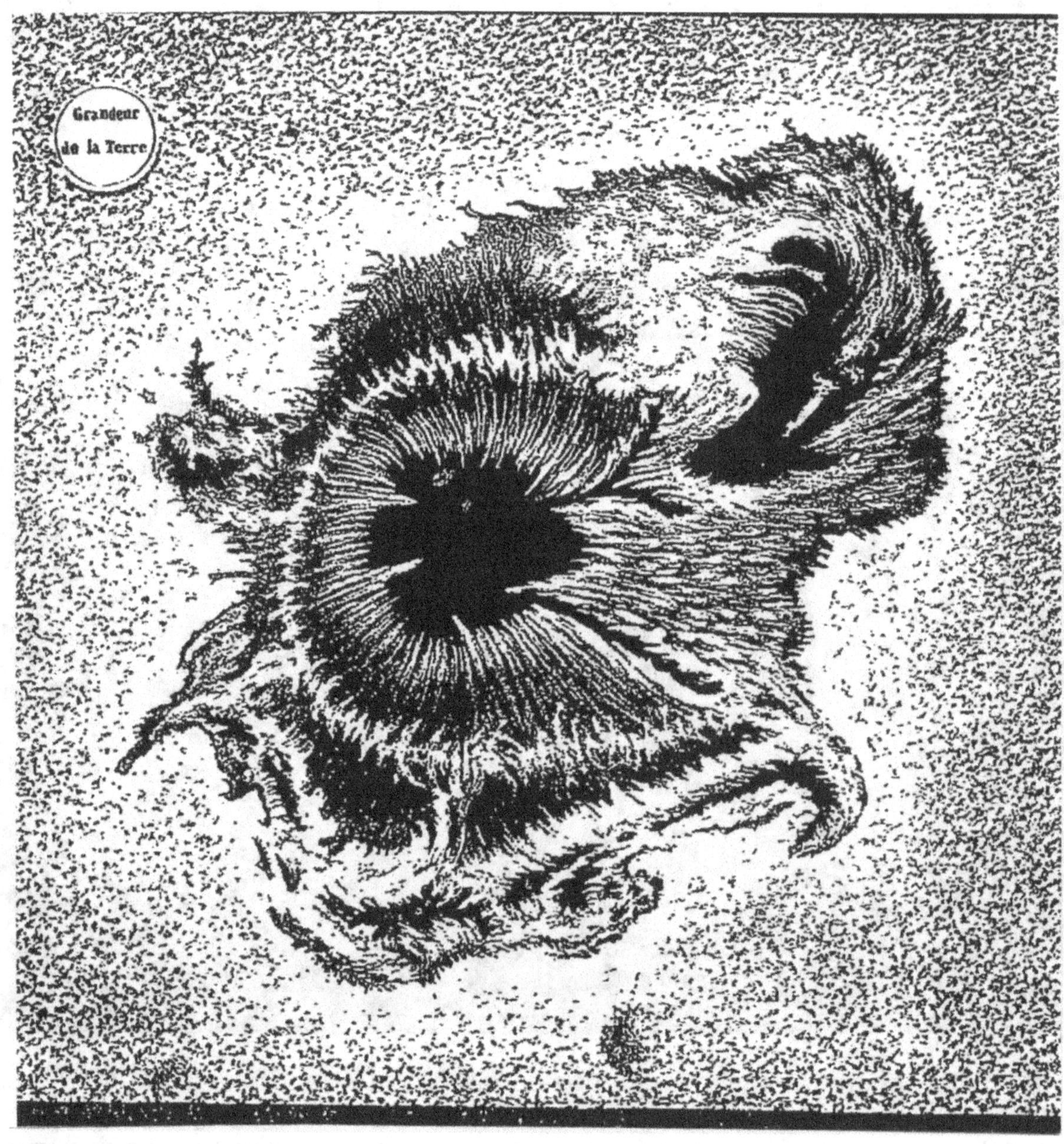

Fig 12. Large sunspots are more often complex groups with multiple orientations as clearly shown on this precision drawing by th eParis/Meudon astronomers, October 14, 1883. This group visible by the nakes eye was seven times the diameter of the earth shown at the top left: "Grandeur de la Terre". From l"Astronomie Populaire" Camille Flamarion, 1891, page 325. [13] Also suggested here is the surrounding "facula", a brighter area than te average surface tempurature.

1. Direct Naked Eye Observations of Sunspots since Antiquity.

The earliest probable observations of sunspots known today are from China as far back as circa 800 BC. In "Solar Observations" Wikipedia reports that the "Book of Changes" mentions "A dow is seen in the Sun" and "A mei is seen in the sun" where "dow" and "mei" would mean darkening or obscuration (based on the context). The earliest surviving record of deliberate sunspots observation dates from 364 BC, based on comments by Chinese astronomer Gan De in a star catalogue. By 28 BC, Chinese astronomers were regularly recording sunspots in official imperial records.

Despite abundant astronomical records in Egyptian and Babylonian "literature", no sunspot observations were evident. The same can be said for Biblical, Sumerian and Roman records: No trace of sunspots so far.
The only clear antique reference to a sunspot in the "Western" world is from Theophastos, disciple of Plato and Aristotle around 300 BC. The Roman silence on the subject is amazing. It takes us 1,100 years after Theophastos to find the next report of sunspots.

At the end of the Dark Age, the only literate people in Western Europe were royalty and monks. Sure enough, it is a Benedictine monk, named Adelmus, who chronicles seeing a large sunspot visible for 8 days in March 807. Adelmus also observed northern lights during that same week and connected the two as related: sunspots create aurorae. A second report by Adelmus is of a sunspot in 813, one year before Charlemagne's death.

In, 1128, the first known image of sunspots is given by monk, later Bishop John of Worcester in his chronicle of the time.
Albert van Helden cited in 1996 the following passage from these chronicles:
"In the third year of Lothar, Emperor of the Romans; in the twenty-eight year of Henry (second), King of the English, on Saturday 8 December, there appeared from the morning right up to the evening two black spheres against the sun". To maximize the chance for middle age scripts scholar to read some of John of Worcester chronicle a full width figure is reproduced here

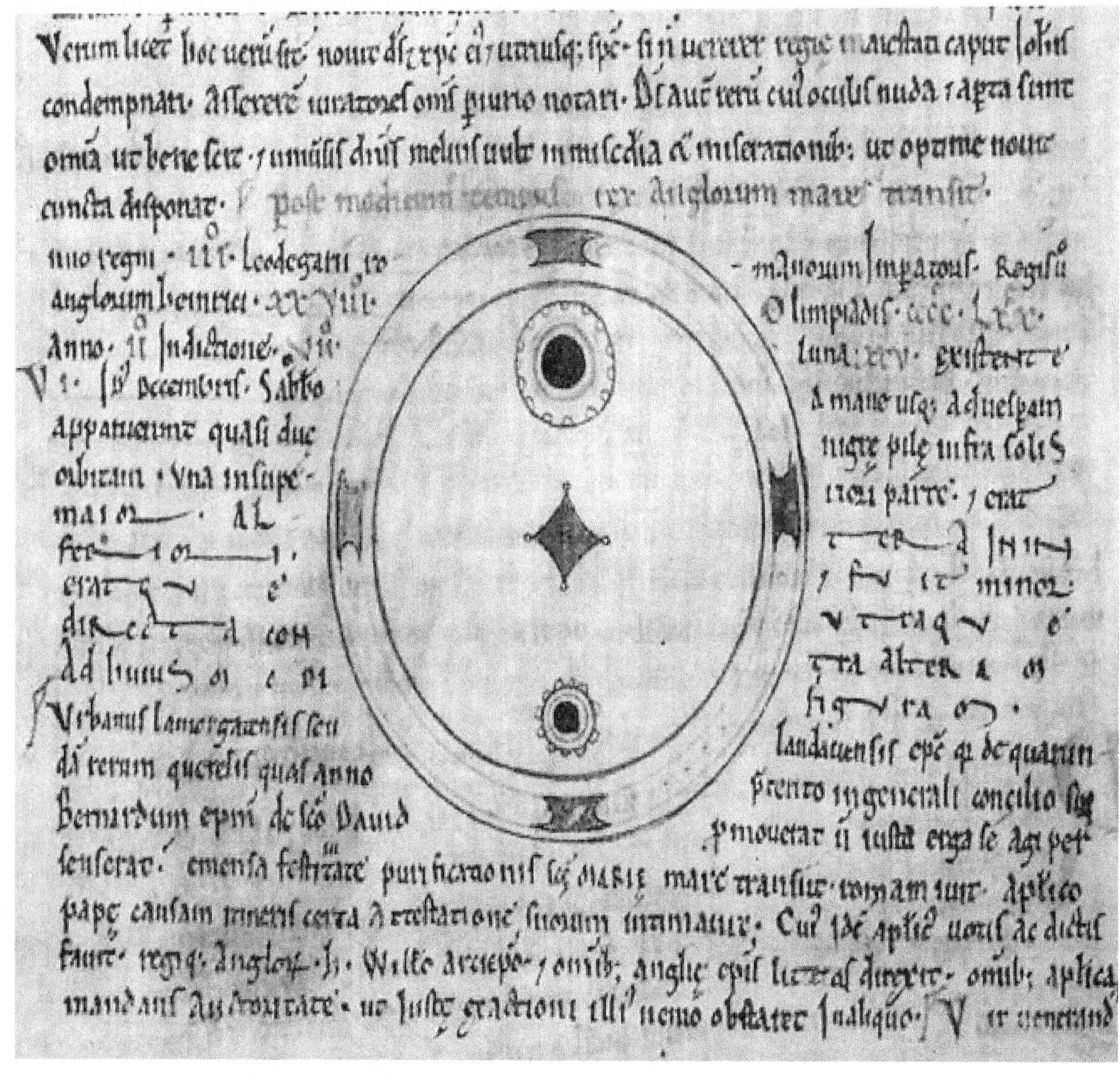

Fig 13 : John of Worcester drawing of two sunspots. Dec 8, 1128

After John of Worcester's chronicle of 1128 AD, there are no known sunspots reports until the late fifteen hundreds.   We will later connect the dots between sunspots and climate. We should have had more reports of sunspots between 1128 and 1250 which was the hottest part of the middle ages as described in Part One.

However, after 1250, and until just before the invention of the telescope in 1608, visible sunspots appeared rarely or not at all.

As reported before, I observed with surprise a large sunspot and smaller surrounding "umbrae" in November 1952. Then, from January to May 1953, in my second candidature year for engineer at the University of Louvain in Belgium, my Professor of Astronomy was the "Chanoine" (enhanced title for a distinguished Roman Catholic Priest) George Lemaitre, the 1927 discoverer of the "primordial atom"[14] later called "the Big Bang" after WWII. During these 3 hours weekly lectures, my opportunity came to ask a question about these sunspots: "Yes, sunspots are well known to exist and so what?" was his approximate reply. It was not until 2006 that I reconnected with sunspots, through the climate literature.

* * * * *

2. 1609. The Telescope immediately Applied by Galileo to Sunspots.

From the history of sunspots viewpoint, the timing of the telescope discovery was exquisite. It triggered intense competition between freshly self-appointed astronomers for priority in all kinds of celestial discoveries including that of sunspots. As a result, we have fairly accurate sunspots numbers from 1610 on, 35 years before the 1645-1700 almost total disappearance of these spots.

The "Verre Kijker" or Far Looker or Longview.
Like today, many people of the Renaissance time had vision problems but no way to correct them. Glass had been produced mainly for wine bottles. This had led to the first translucent windows made by assembling wine bottle bottoms together and joining them with lead beads as earlier in middle age cathedral windows. Better selection of raw materials improved glass making to the point of colorless glass produced for "spectacles". This became, by 1600, good business to correct vision problems for people with money. In most cases, it was a simple "monocle". Convex or concave lenses were made by smooth polishing of thick initial glass bodies. This required ingenuity and tough work.

The traditional story[15] is that a spectacle maker named Hans Lipperschey handled two "spectacles", one concave and one convex in line with each other and saw the steeple of a nearby church enlarged. This happened in Middleburg, Holland in 1608. To lock the two spectacles or lenses in their proper position for sharpness of view, he inserted them at the two ends of a tube. He and his associate spectacle maker, Zacharias Hanssen thus created the first "longview" which they called "Kijker" or looker or viewer and filed for patent protection, October 1608. In a conflicting claim, credit should be attributed to James Metius of Alkmaar but we have no other information on that issue.

The speed at which the telescope spread across Western Europe is barely credible for the time. Indeed, by July 1609, both Thomas Harriot in England and Galileo Galilei in Italy started to use the magnificent Dutch invention for astronomy. Harriot was apparently the first to observe details of the moon surface, while Galileo was the first mid-fall 1609 to follow a sunspot across the surface on the sun. It was a feverish competition in lens polishing, understanding what made greater magnification and selecting the subject of "looking" to be the first to discover planetary details.
Today, the story of this frenetic competition evolves with new documents being pulled out and other ones forgotten. A key example of convenient neglect is the publication in March 1610 by Galileo Galilei of "Siderius Nuncius".[16] In this forgotten manuscript Galileo describes Saturn's rings, several satellites of Jupiter and a large sunspot traveling from west to east on the sun, becoming elliptic before disappearing on the other side. It is over a year later in 1611, that Galileo goes to Rome with the story of his sunspot. Encouraged by the success he is receiving, he then goes too far for the Vatican by publishing in 1613 his "Letters on the sunspots". For Galileo, the movement of the sunspots on the surface of the sun is strong evidence that the heliocentric concept of the world by Bishop Copernicus is the correct one. And as is well known, Galileo is condemned for life to seclusion in his Florence home.

This first observation by Galileo in the fall of 1609 of a sunspot moving from West to East on the surface of the sun gives us the opportunity to show such a spot close to the eastern edge of the sun by Flamarion.

Fig 14 "Spot arriving at the edge of the sun". This drawing by the Paris/Meudon astronomers clearly suggests a protuberance around the dark central area. From l'Astronomie Populaire, by Camille Flamarion, 1891, page 331.

Flamarion comments: "The spots are usually surrounded by very brilliant areas named "faculae" in the solar literature. These are swellings of the photosphere clearly seen when the spots approach the solar rim as in the figure above. These regions are thus the sites of considerable agitation of which surface surpasses greatly that of the spot itself".

* * * * *

3. 1610-1630. Priority Battles Promote Systematic Sunspot Recording.

Newly minted astronomers sprang out of nowhere from 1609 on and revolutionized the knowledge of celestial bodies thanks to fast improving telescopes. This was late Renaissance with the birth of science by experiment and systematic observation over blind faith and old beliefs. By 1613, there was an additional incentive: Notoriety for being first! A brief summary is given of the best known names of the time as of today:

<u>Thomas Harriot.</u> England. Thought he was the first to observe sunspots in late 1610 but Galileo had done it in 1609 and published in March 1610. He, however, had his first telescope in July 1609 about 2 months before Galileo but observed the moon with it at night.

<u>Johannes and David Fabricius.</u> Leipzig, East Germany. They published a description of sunspots in June 1611, using the camera obscura projecting the solar disk on paper.

<u>Christopher Scheiner</u>. Ingolstad, Rome. German-born Jesuit at the Vatican. Made his own good telescope with camera obscura for sunspots and wrote a voluminous account of observations. That book published in 1630 is entitled "Rosa Ursina". He fought Galileo for priority and took sides with the Pope on astronomy. His book has been essential for reconstruction of sunspots counts, much later by Rudolf Wolf and others.

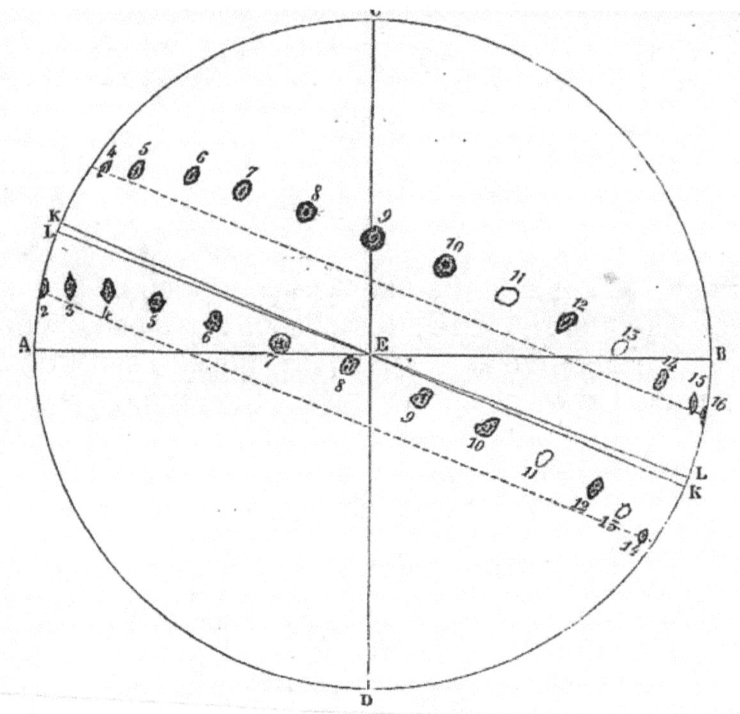

Fig 15. A 1627 drawing by Christopher Scheiner showing for the first time the rotation of the sun revealed by two sunspots observed once a day for 2 weeks except days 11 and 13 probably cloudy. Scheiner got in serious trouble with his superior for having observed "imperfections" on the sun. Flamarion, 1891.

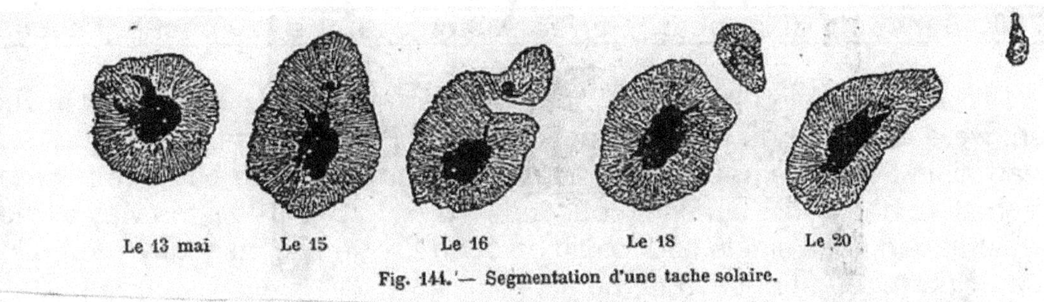

Fig 16. Sunspots evolve with time. Here is an example of progressive segmentation of a spot over 7 days in the month of May sometime in the 1880ies. From l'Astronomie Populaire 1891. Flamarion p.328.

Galileo Galilei. Padua, Florence. Traveled to Venice in May/June 1609 where he was told of the Dutch dual spectacle "Kijker" at a local fair. Back to Padua, he immediately started polishing lenses to make his first, crude, 3X telescope. By September, his understanding of lens curvature and improved polishing techniques allowed him to manufacture a 33X telescope which he applied immediately to astronomy. Observing a sunspot moving, over one week or two, from the left to the right of the sun and becoming elliptic before disappearing on the other side was accomplished among other nightly discoveries around Saturn and Jupiter. That was Fall 1609. In march 1610, he published his manuscript "Siderius Nuncius" in which the above observations were described. This proves Galileo Galilei's priority on sunspots observations.

By 1630, the priority controversy had subsided but not the faithful observation and recording of sunspot numbers.

The Birth of Observatories around Western Europe and the World.

Already with a 33X magnification, the need to stabilize the telescope is quickly apparent. Next, the tubes became more solid with rolled up sheets of armor or brass. Telescopes had to be installed on a firm rotating pivot to stabilize the view and support the fast increasing weight. This led quickly to national observatories in Switzerland, England, France, Germany, Denmark, etc.. The technology of telescopes grew by leaps and bonds and the progress continues today. Thinner, dryer atmospheres of high mountain peaks led to Mount Palomar and Mauna Kea as recent locations for the best telescopes of the 20th Century. Today it is from space, above earth's atmosphere that the most precise images of our universe are reaching us.

* * * * *

4. 1645-1700. Sunspots disappear. Perfect Match with the Deepest Little Ice Age.

Barely in business observing, counting and recording sunspots since 1609, budding astronomers were all astounded in 1645 and beyond. For the first time, they all agreed, sunspots had all but disappeared. From numbering roughly 10 to 75 depending on the year between 1610 and 1645, sunspot counts dwindled to 0 to 10 all the way to the year 1700. Even stranger, not a single spot could be seen for many consecutive years. It was not an issue of lack of effort observing the sun.

The following figure composed 300 years later by Dr. John A. Eddy (see pts 10 and 11) reconstructs what early astronomers reported during the "spotless" 1645-1715 period later called the Maunder Minimum.

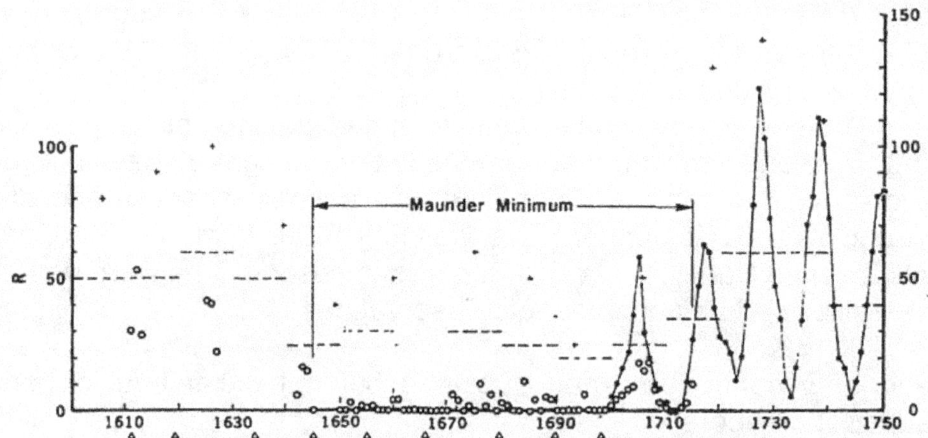

Fig. 9. Estimated annual mean sunspot numbers, from 1610 to 1750: open circles are data from Table 1; connected, closed circles are from Waldmeier (*3*); dashed lines (decade estimates) and crosses (peak estimates) are from Schove (*8–11*); triangles are Wolf's estimated dates of maxima for an assumed 11.1-year solar cycle (*3, 6*).

Fig. 17

I made a point of including Eddy's description of the Maunder Minimum figure as is, to show the complexity and apparent contradictions in some of his sources. However, I am surprised that a wealth of sunspots data are not included from the 1630 book "Rosa Ursina" by Christopher Scheiner. At least from 1612 to 1629, there would be clearer yearly sunspots counts, very useful to appreciate the details before 1645.

Doesn't it also appear strange to the reader that Eddy elongates his "Maunder Minimum" to 1715? In fact, it is clear from the diagram that the very low sunspot count ends in 1700 on the dot. So, to this author the Maunder Minimum is 55 year long, 1645-1700.

Its effect on climate though was prolonged by two to three decades. It was hysteresis or continuing cooling effect by advancing glaciation that was poorly recorded at the time. In Part One, we elaborate on the 1700 to 1750 delay in warming.

Sunspots Control Climate

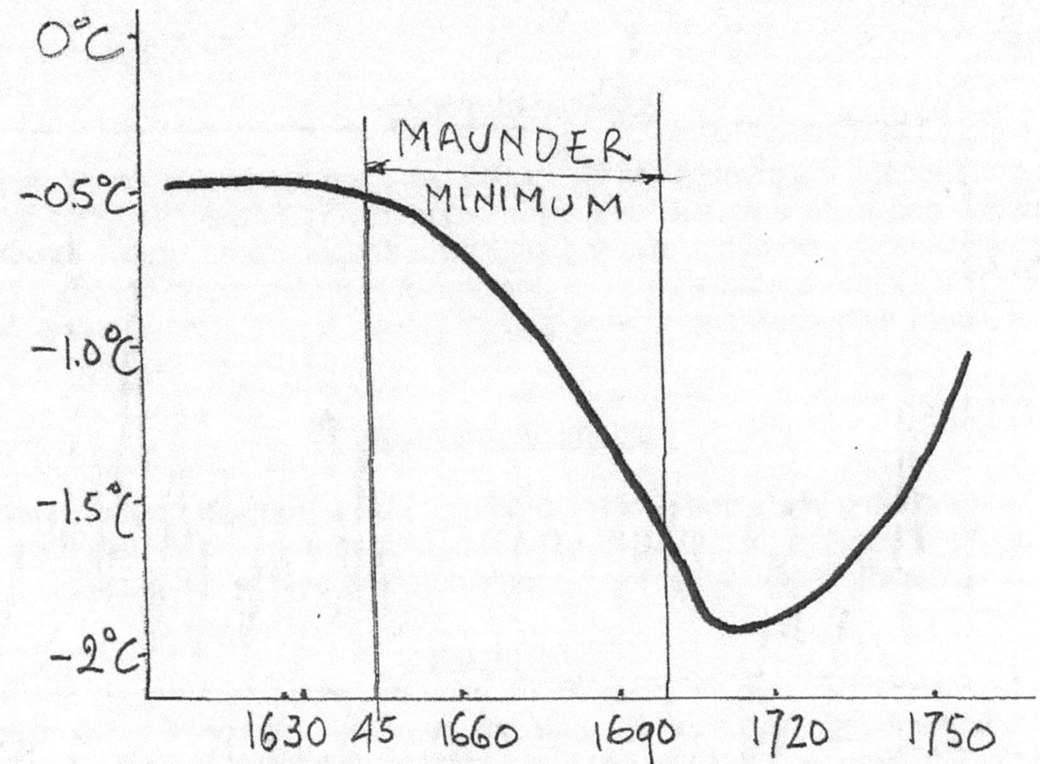

Fig 18. The author's perceived – 2° C drop during the Maunder Minimum.

The deepest, -2°C "Little Ice Age" coincides with a near total sunspots disappearance The Maunder Minimum.

* * * * *

5. 1700-1789. Strong Resurgence of Sunspots. The Onset of our Little Warm Age.

The climate of the entire 18th Century is dominated by a resurgence of solar activity clearly shown in fig.19 by the sunspots count.

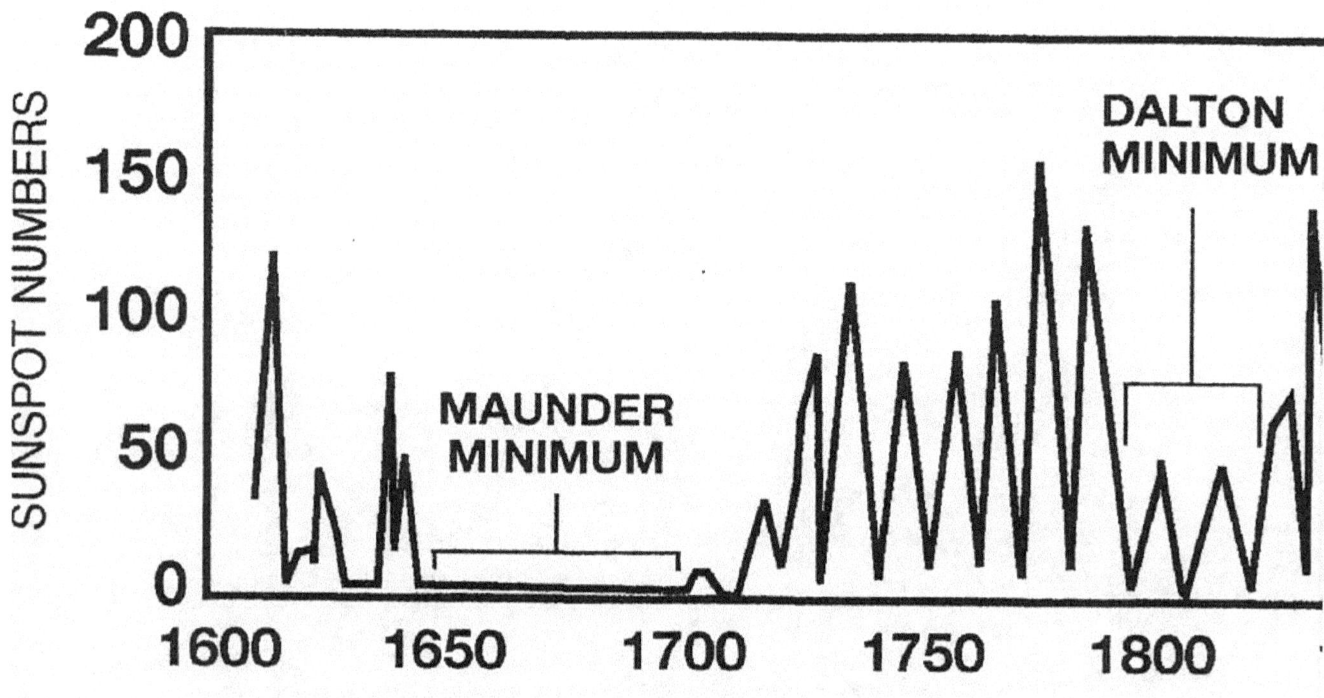

Fig.19. This schematic diagram of sunspot count versus time clearly shows:

 1) The descending trend from 1610 to 1645 recorded by early astronomers.
 2) The 55 year near total absence of spots now called the Maunder Minimum.
 3) The 90 year resurgence of the 18th century, strong enough to cause substantial warming at least
       100 years before the Industrial Revolution. Ambler agrees with the 55 year Maunder
From the book: "Don't sell your coat"[17] by Harold Ambler, page 159

As we pointed out in Part One, after 55 years of cooling, the climate had hit bottom, more than 2^0 C below neutral. It took some 50 years of growing sunspot activity for temperatures to recover back to neutral around 1750. After 1750, as we have demonstrated in Part One, the Glacier of Glacier Bay retreated 5 miles back to the shore by 1794. The resurgence of sunspots throughout the 18th century clearly shows the cause of that early warming, 100 years before the industrial revolution.

* * * * *

6. 1795-1825 The First Dalton Minimum.

After a healthy sunspots activity throughout the eighteen century a marked reduction in sunspots count appeared at the end of cycle 4 which peaked in 1787-88.

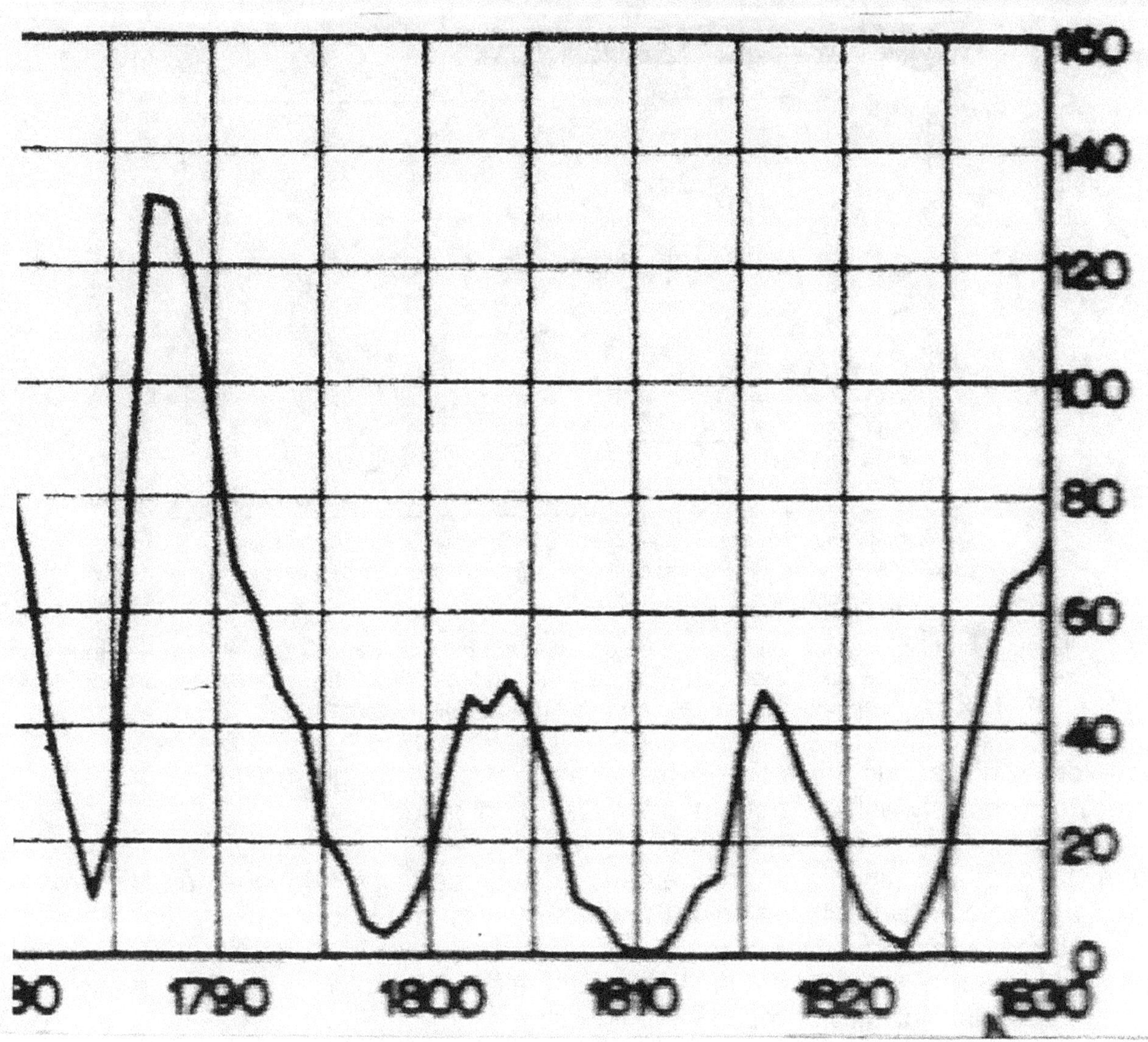

Fig 20 . Annual Mean Sunspot Number. Enlarged from M. Waldmeier[18]

After the peak #4 of 1787-88 with about 130 sunspots the descending curve is abnormally elongated with a middle inflection which we will analyze below. Then double peak #5 in 1803-1804 barely reaches 43-45 sunspots. The following bottom from 1809 to 1812 shows zero sunspots for 3 years. Cycle #6 then peaks also at 44 sunspots in 1817 followed by the last Dalton bottom in 1823. The following cycle #7 shows the end of Dalton, peaking again at 80 sunspots.

Let's now go back to the uncommon descent of cycle #4 in the following figure.   Jan Alvestad of the Solar Influence Data Analysis Center in Brussels, Belgium, analyzes in detail what is now called 4 prime or even better the "lost cycle":[19]

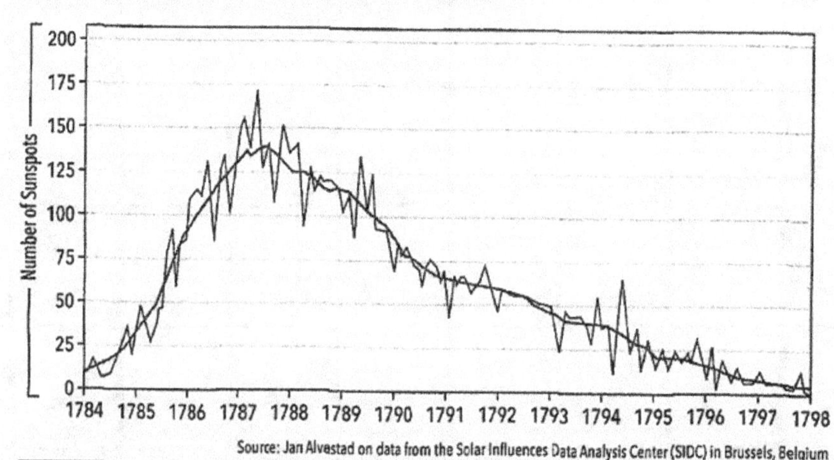

Figure 2-2: Note the period from 1793 to 1798. This is the period identified by Usoskin et al. as 4 Prime, the "Lost Cycle." I concur with their findings and have used 4 Prime as my start to the Dalton Minimum.

Fig 21. Cycle 4 and its long final tail in great detail. The anomaly is immediately apparent at the onset of the descent. It's angle is weaker than that of the ascent. It looks like there is an aborted cycle between 1791 and 1796. It has been called "4Prime" or the "lost cycle" Usoskin et al.[20]

Napoleon's disastrous invasion of Russia and retreat through the winter of 1812-1813 coincides precisely with the bottom of the Dalton minimum.

John Dalton, the distinguished mathematician and chemist of Manchester, UK. was also a self appointed meteorologist. In 1787, he started an assiduous observation of weather with self made instruments and maintained this diary for 57 years with 200,000 entries. His first separate publication in 1793 was titled "Meteorological Observations and Essays". It did not attract much attention at the time. The 57 year diary is now an excellent source of climatic evidence for the period 1787-1840, covering the Dalton cooling and recovery.
In today's media-standard climatology, the Dalton minimum is the last part of the multiple mid-millennium Little Ice Ages. In effect, as we have demonstrated in point #5, the official consensus is to ignore the strong warming of the 18th Century.

If and when climatologists will take their North Star from sunspots cycles, this aberrance of 1850 as the end of Little Ice Ages will evaporate. Our current Mini Warm Age wil be then recognized as starting in 1750.

* * * * *

In points 3 and 4, we underlined the competitive recording of sunspots since the advent of the telescope. However, for well over 200 years, no one got the idea of synthesizing those records and organizing them into a linear history. As often the case, it is an amateur who triggered the move to a comprehensive picture.

(Samuel) Heinrich Schwabe. (1789-1875) Dessau- Rosslau, Germany.

Fig22. Wikipedia photo of Heinrich Schwabe

Educated as a pharmacist at the Humbolt University of Berlin, Schwabe only turned his attention to sunspots at 37. We don't know what focused him on sunspots. We do know however that, equipped with his own telescope, he faithfully counted sunspots almost daily for 17 years, 1826-1843. That was enough for him to be the first to detect a cyclical behavior:"There are about 10 years (later adjusted to 11) between two maxima." Heinrich Schwabe had, in fact, captured the end of the Dalton minimum described in point 6, the rise to cycle 7, the drop almost to zero and the high peak of cycle 8 to 140 sunspots. This was the object of a brief publication in 1843.

The sunspots cycles are now called the Schwabe cycles.

Johann Rudolf Wolf. (1816-1893) Born near Zurich, Switzerland.
His family can be traced back 500 years. His father Johann was a Minister of the Church. Rudolf first graduated from Zurich's University then perfected his education as an astronomer in Vienna and Berlin.

Fig 23. Wikipedia photo of Johann Rudolf Wolf

Soon directing the Swiss Observatory still in Bern at the time, Wolf got very interested in Heinrich Schwabe's 1843 publication on sunspots cycles. That triggered a decade long research into the old documents on sunspots. First, Wolf tested the validity of Schwabe's cycles by extending back before 1826 the observations he could find. Frustrated by the lack of uniformity in old data, Wolf then organized, in 1848, a "Pan-European" effort at standardizing sunspots counts and records in National Observatories. He moved the Swiss Observatory from Bern to Zurich.

Most importantly, Wolf dug deep into many old manuscripts and succeeded in reconstructing a fairly accurate picture of sunspots cycles back to 1700, see point #5. He is thus the one who established the length of the cycles to be eleven years on average. Beyond 1700, however Wolf became suspicious of the data he could find and hardly believed the absence of spots for so many years. Rudolf Wolf thus did not acknowledge the 1645- 1700 minimum. That was to come next.

Friedrich Wilhelm Gustav Spörer. 1822-1895. Berlin ,Giessen.
Educated in mathematics and astronomy at the Humbolt University in Berlin, Doctor Spörer dedicated a significant part of his later career to sunspots research. The result:

Fig 24
"The 70 year Sunspot Minimum". Spörer went further than Wolf before 1700 and was first to identify a long sunspots minimum.

In papers published in 1887 and 1889, Spörer called attention to "a 70 year period ending about 1716, when there was a remarkable interruption in the ordinary course of the sunspot cycle and an almost total absence of spots".

At the Berlin Observatory, Spörer was studying the distribution of spots with latitude and had found evidence that Northern and Southern frequencies were not always balanced on the visible face of the sun. To check his personal telescopic observations, Spörer consulted historic documents including Wolf's and was surprised at what he found in the data of the late 17th and early 18th centuries. He then reconstructed the low sunspot event mentioned here above. He died soon after, in 1895.

In fact, Spörer was the true discoverer of the Maunder minimum.[2] It should have been the Spörer minimum. To recognize his key contribution, he now has his minimum, from 1460 to 1550. Today's sunspots oriented climatologists have proposed 3 mid-millennium Little Ice Ages: The Wolf (1300-1370), the Spörer (1460-1550) and the Maunder (1645-1715).

Edward Walter MAUNDER. 1851-1928. London, Greenwich.
A British astronomer, Maunder became photographic and spectrographic assistant at the Royal Greenwich Observatory near London.

Fig.25 Wikipedia photos of Edward Maunder

By 1890, E.W. Maunder was promoted Superintendent of the Solar Department, Greenwich Observatory. Having read the 1887 and 89 Spörer papers he took up the case. He summarized the Spörer message in an 1890 paper at the Royal Astronomical Society in London.

Then based on his own research in old documents, Maunder published in 1894 a major article titled "A Prolonged Sunspot Minimum".

The unexpected lack of response to this outstanding contribution deeply affected the solar astronomer. After almost 30 years of waiting and in desperation, Maunder tried again in 1922. He upgraded the previous paper with new data but kept the same title. The most critical news was a claim by Agnes Clark that the prolonged sunspot minimum was also a prolonged absence of Aurora Borealis.

That second paper fell flat as well and Maunder died six years later still baffled by the silence of the astronomical society of Europe.

It looked like the magic relay race started by Schwabe and taken up by Wolf and Spörer had collapsed after Maunder. Nobody to take up the sunspot challenge. This silence actually deepened for 54 years after Maunder's second paper, until 1976. (point 9). After the discovery of the magnetic nature of sunspots, see point 8, during the 1910-1920 decade, Maunder dabbled into the subject but did not contribute any breakthrough.

Some main results from Maunder's research are summarized hereunder, relayed by Dr. Eddy: [2]

•   The Philosophical Transactions of the Royal Society reported in 1671:
"At Paris, the excellent Signor Cassini has lately detected Spots in the sun of which none had been seen in these many years that we know of."
Cassini's own description reads as follows: " It is now about 20 years since astronomers have seen any considerable spots on the sun before that time, since the invention of the telescope, they had observed them from time to time". Cassini mentioned Picard, an other French astronomer who was also pleased at finding a sunspot after 10 years.

•   In 1684, the Royal Astronomer at Greenwich, Flamsteed reported seeing one sunspot and commented:

"These appearances (of sunspots) however frequent in the days of Scheiner and Galileo, have been so rare of late that this is the only one I have seen in his face since December 1676".

•   After all, neither Spörer nor Maunder had "discovered" the Sunspot Minimum.
It had been reported matter of factly in early Astronomy books such as "Astronomie", 3 volumes written by Joseph Jerome Le Français de La Lande in 1792 in which he gave dates and details of the prolonged sunspot minimum.   William Herschel also mentioned the minimum but based on La Lande's books.

In conclusion, a summary of the 5 Spörer and Maunder articles can read:

  1 - From 1645 to 1715, very few sunspots were detected.
  2 - From 1672 to 1704, not a single spot was seen in the Northern hemisphere of the
            sun.
  3 - From 1645 to 1705, no more than one Spot Group could be seen at one time.
  4 - From 1645 to 1715, only a "handful" of spots were seen, mostly single, only at
            lower latitudes and lasting one rotation or less (27 days).
  5 - From 1645 to 1715, the total number of spots was less than that of the peak of
            a normal cycle.

From all the period's data, I venture to suggest that no more than One Spot per year, on average, or 55 spots grand total were observable from 1645 to 1700.

In 1976, Eddy titled his article: "The Maunder Minimum" and it stuck! [2]

* * * * *

8. 1896-1908. Zeeman and Hale Establish: Sunspots Are Strongly Magnetic.

Born in 1865 at Zonnemaire on the island of Schouwen, Zeeland, the Netherlands, Pieter Zeeman studied Physics at the University of Leiden where he taught from 1890 to 1900.

Then promoted Professor of Physics at the University of Amsterdam, he received the Nobel Prize of Physics in 1902, shared with Hendrich Anton Lorenz. In 1908 he was named Director of the Physical Institute of Amsterdam. Pieter Zeeman died in Amsterdam in 1943.

The Zeeman Effect: Splitting Spectral Lines by Magnetic Field.

Michael Faraday had discovered that the polarization of light was influenced by magnetic fields but could not specify clearly the interaction. Zeeman investtigated this "Faraday Effect" with his famous 1896 experiment: Placing a sodium source in a flame between the poles of a strong magnet, Zeeman observed the splitting of the strong 5890-5896 Angstrom yellow spectral line into multiple components in a spectrograph. His colleague Hendrich Lorenz theorized that the splitting occurred in 3 (a triple) and the angle of widening was proportioned to the intensity of the magnetic field. His further theorizing was proven inaccurate later and required Quantum energy levels being developed at the time by Max Plank to be completely understood.

The Zeeman effect has many applications in today's medical and electronic fields.

Fig 26

Pieter Zeeman

Dutch physicist

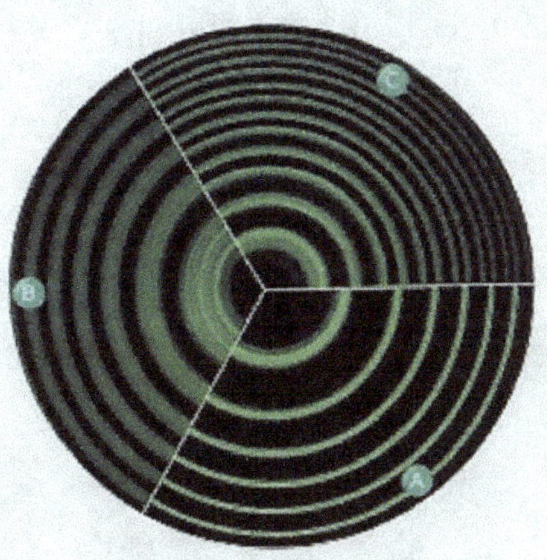

The spectral lines of mercury vapor lamp at wavelength 546.1 nm, showing anomalous Zeeman effect. (A) Without magnetic field. (B) With magnetic field, spectral lines split as transverse Zeeman effect. (C) With magnetic field, split as longitudinal Zeeman effect. The spectral lines were obtained using a Fabry–Pérot interferometer.

Fig 27

Fig. 26 & 27

<u>George Ellery Hale. 1868-1938. Brilliant American Astrophysicist.</u>

George's father, William Ellery Hale was a very wealthy businessman who strongly supported his son's ambitious goals in astronomy. They lived in Chicago but George was an undergraduate student at MIT in 1889. From those studies, it is probable that Hale knew of the 1870 invention of the spectrohelioscope by C.A. Young. In any case, in 1889, in recess in Chicago, he received out of the blue, during a streetcar ride, the best idea of his life: the **spectroheliograph**. He would transform an existing telescope into an enlarging and protected spectrograph to analyze in monochromatic light the chemical composition of sun and stars. Hale was in such a hurry that he never graduated from MIT as PhD although he certainly deserved it. By 1891, his first spectroheliograph had been installed on his backyard family telescope, the "Kenwood Observatory" at 4545 South Drexel Avenue.

Young, brilliant, and ambitious, Hale was the ideal collaborator with Harper in founding the University's astrophysical research program.

Fig 28

Forty-inch telescope, Yerkes Observatory.

Fig 29

At the University of Chicago from 1891 Hale was permitted to focus all his energies on the new observatory to be built at Williams Bay along Lake Geneva, Wisconsin, called:The Charles T. Yerkes Observatory. It opened in 1897. Its 40 inch lens represented the best magnification to date and allowed new star parallax measurements. Equipped by 1903 with the second, improved spectroheliograph, this telescope made the first photographs of the sun's blue hydrogen lines.

Fig. 28 & 29

52

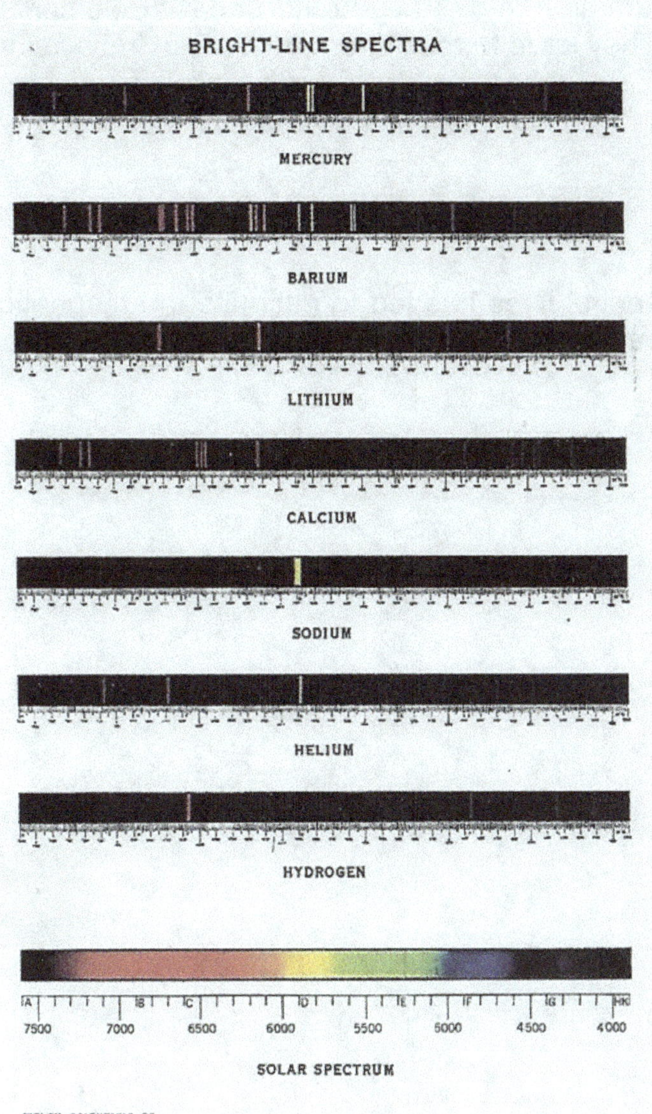

WELCH SCIENTIFIC CO.

Reproductions of the bright-line spectra of selected elements found on the sun, and of the solar spectrum with principal Fraunhofer lines A, B, C, D, E, F, G, H, K. A and B are caused by terrestrial oxygen (O_2); C and F belong to hydrogen; D represents sodium; E, G, H, K indicate calcium mainly. Also shown are spectra for mercury and barium

Fig 30

The main hydrogen red line had to wait for better photo film emulsion sensitivity. It is only at the 60-inch Mount Wilson Observatory that, in 1908, George Hale discovered and photographed the splitting of the red hydrogen spectral line. He thus confirmed the Zeeman effect of sunspots.

Hale's team also discovered the polarity of sunspots and established the presence of **many pairs of spots** with the opposite polarity.

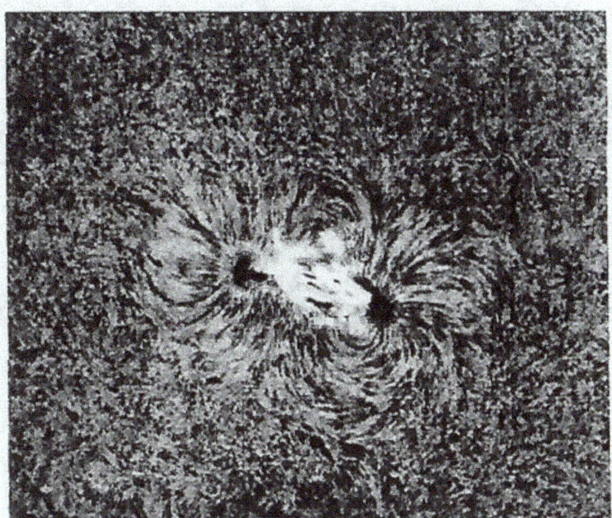

HYDROGEN CLOUDS NEAR SUNSPOTS

Hydrogen clouds surrounding a large, magnetically bipolar group of sunspots, photographed at the Mt. Wilson observatory with the 13-ft. spectroheliograph using the red hydrogen line (Hα). The forms of such clouds resemble those of iron filings over a bipolar magnet or flow lines around double vortices rotating in opposite directions

Fig 31 From Encyclopedia Brittanica 1972

Fig 30 & 31

Hale also detected the presence of other elements than hydrogen and helium on the surface of the sun, including calcium, barium, iron, etc. These other, heavier elements cannot result from fusion beyond helium. They have to result from myriads of impacts of asteroids over the sun's 5 billion year life.

Today, the angle of splitting of iron, Fe spectral lines has led to estimate the magnetic field near the center of large sunspots to reach 3,500 times the one gauss magnetic field of our planet or 3,500 gauss.

* * * * *

<u>**9. 1930-2000. Peak Sunspot Activity Coincides with CO$_2$ Rise in the Air.**</u>

After the first Dalton minimum we left sunspot evolution in the 19th and 20th Century for later. Here it is in the following diagram showing:

•   Strong mid 1800 cycles.
•   A secondary but distinct weakening from 1880 to 1930 which I venture to call the Industrial Minimum.
•   The peak activity from about 1930 to 2000.

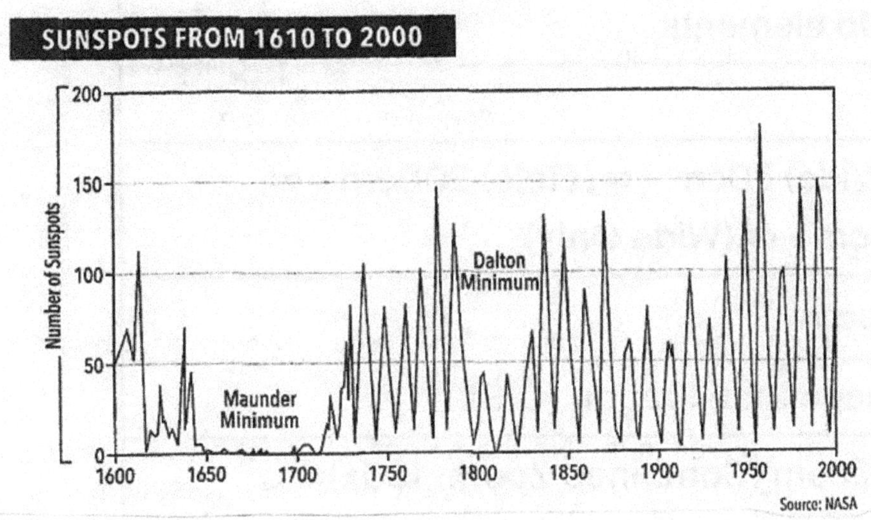

Fig. 32. A comprehensive, simplified diagram developed recently by NASA showing the 11 year Schwabe cycles. The last peak in 1998 is peak #22. The number one is, unfortunately, much too late for reality: it is the first one after the 1750 peak, the 1761 peak. This diagram also confirms the strong but ignored 18th century activity. From "Dark Winter" by John L. Casey.[22]

Precise carbon dioxide analysis in our atmosphere started after WWII. One of the best known today was initiated by Charles Keeling around 1953 on top of Mauna Loa, the second highest extinct volcano on the big Hawaiian island. Being located in the middle of the Pacific Ocean and despite occasional volcanic eruptions emitting CO$_2$, the Keeling station is ideal for reflecting a fully homogenized global atmosphere. Fig.33 taken from my 2014 publication shows the diagram ending in 2011 which fully covers the need of point 9. Today, however, the continuing asymptotic or exponential shape of the curve reaches around 420ppm in 2020.

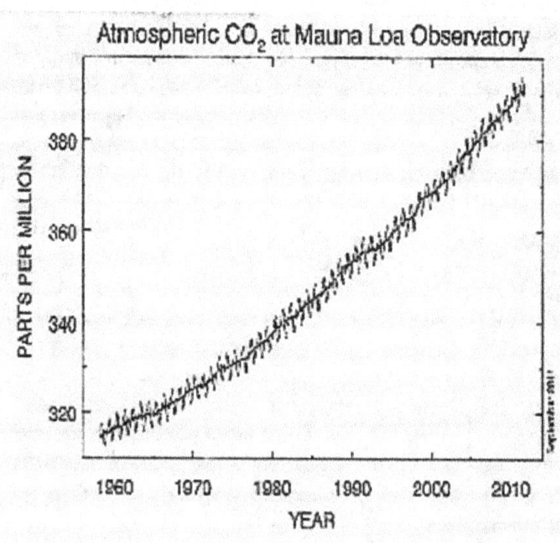

Fig 33 Detailed evolution of CO$_2$ as a function of time in the Atmosphere. The Keeling Curve. Sept. 2011

Every Spring-Summer period in the Northern hemisphere shows a significant 1-2ppm decline in CO$_2$ content. This is followed by a 2 to 4ppm resurgence in Fall-Winter of the North because the total CO$_2$ absorbing vegetation of the Southern hemisphere is much smaller and incapable of compensating for the human production of CO$_2$.
By now superposing in mind the 1930-2000 sunspot surge and the ascending CO$_2$ curve, one can be confused on which factor impacts our current climate change. It is only by examining what happens to both curves after 2000 that the problem can be solved as we will elaborate in point 11.

As we have developed since Point 1. an extensive story of sunspots, let us examine the same for Carbon Dioxide, CO$_2$, as succinctly as we can:

Two billion years ago, the earth's atmosphere was essentially Nitrogen (N$_2$) and Carbon Dioxide (CO$_2$) on a dry base. Minor constituents were Sulfur Dioxide (SO$_2$) inherited from decreasing but still higher than today, volcanic activity and noble gasses as today, Argon, etc.., about 1%.

56

In Fig. 34, the evolution of N_2 and CO_2 from 2 billion years ago to today is summarized. The reality was not that smooth but the inexorable trend is correct: When Chlorophyl based life started, it split the CO_2 molecule into C which is absorbed in plants and O_2 entering the atmosphere. For hundreds of millions of years, that free oxygen first oxidized enormous tonnages of elemental iron and produced huge iron ore deposits and sequences of sandstones and iron oxide called Itabirites in South Africa and Australia for example.

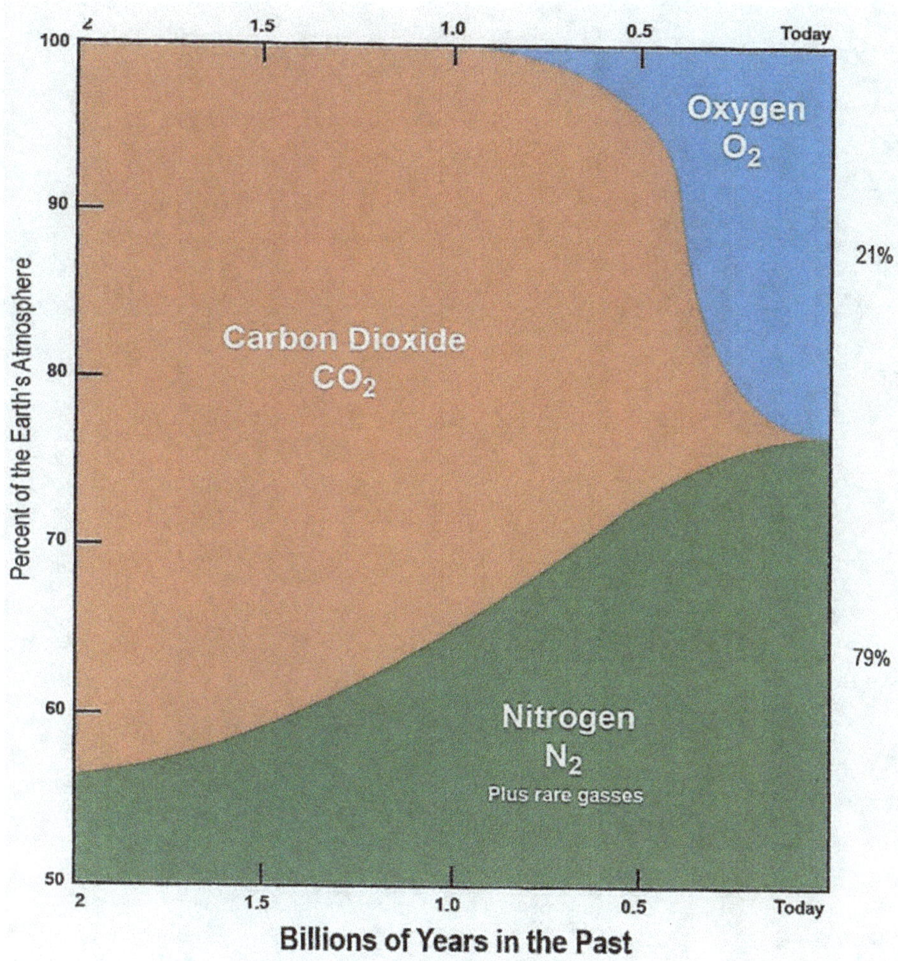

Fig 34. Schematic Evolution of Earth's Atmospheric Composition on a Dry Air Basis over the last 2 Billion years. Ref.[1]

It is only after all that free iron was oxidized to Fe_2O_3 that oxygen began its rise in our paleo-atmosphere. Today we can affirm that every single molecule of oxygen in the air came from a molecule of carbon dioxide split over the eons by the chlorophyll reaction of vegetal life.

However, there was far more CO_2 in the atmosphere than that split to carbon and oxygen by vegetal life. Trillions of tons of Limestone (**CaCO$_3$**) and dolomitic lime stone (Ca, MgCO$_3$) are the result of CO_2 being precipitated in oceans by Calcium and Magnesium. Shells and coral reefs are just a small example. All limestone layers seen in geology contain about 44% of CO_2 by weight.

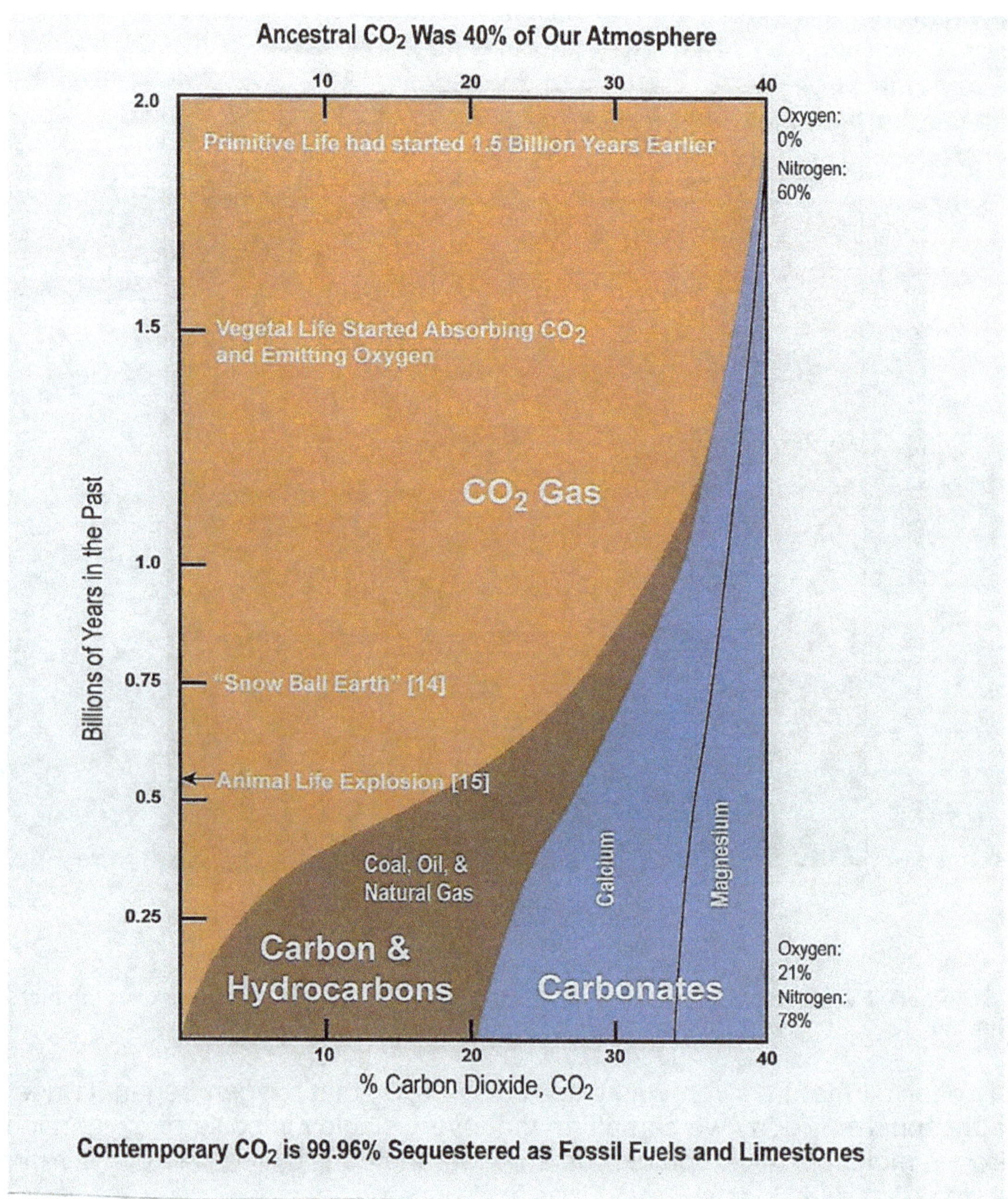

Fig 35. Shows the schematic evolution over 2 Billion years of the CO_2 component in our atmosphere from over 40% to about 0.04% today.   Ref. [1]

To arrive at 40% CO_2, first 21% Oxygen today replaces the same 21% of CO_2 in the past. Then, we omitted the oxygen oxidizing iron. Then, we assumed that the same amount of CO_2 went to limestones, thus another 20%. But it could be much higher after estimate of lime plus dololimes weight has been obtained. Carbon dioxide in our early paleoatmosphere could have been as high as 70-75% of the total with nitrogen volume unchanged, down in percentage to 25-30% on a dry basis.

The net result of this paleoatmosphere discussion for today's climate application is as follows:

If 0.02% CO_2 can move our average temperature up by 1°C, then 40-70% CO_2 would have caused a "run away greenhouse gas effect" one thousand times higher, prohibiting all possible life on earth. Foreseeing this back in the 60ies and 70ies, Carl Sagan introduced a "Deus ex Machina": The young sun must have been "a lot" colder! We will review this in further depth in Part Three but it shows how shaky the CO_2 theory appears to geologists.

* * * * *

John Allen Eddy

John A. Eddy

Before we proceed, let us place this 1976 breakthrough publication in Science titled "The Maunder Minimum" in its time:
There was no trace of the CO_2 greenhouse theory in the press or on the airwaves. Ironically, it was the time of headlines predicting an imminent slide into the next major ice age! Despite its high connecting value with climate, the message barely hit the big headlines because there was no talk of global warming yet.

Fig. 36

Dr. John Allen, "Jack" Eddy. (1931-2009) Born in Pawnee City, Nebraska.

From humble origin in a small prairie town, Jack Eddy patiently broke through financial and political barriers to reach unpredictable prominence in solar science. He loved stars and astronomy, particularly the sun but had no idea, at first, how to morph it into a career. It took local help, Senator Ken Wherry, to leverage that gift through the US Navy and, finally at the University of Colorado where, in 1961, he obtained his PhD with a thesis titled "The Stratospheric Solar Aureola".

One peculiar tendency distinguished Jack from all his academic colleagues and got him repeatedly in political trouble:

<u>History as a major part of Science</u>

Abandoning the fashion of sophisticated instrumental research like that of his PhD thesis, Jack took to old historic records to connect the dots and draw breakthrough conclusions on sunspots and climate.

The famous "Maunder Minimum" paper was the result of such unorthodox approach to scientific progress. It was sorely needed for the future of climate science and became one of its cornerstones today.

The style of Dr Eddy's paper belies expectations of definite statements. It is highly analytical, humble, even hesitant for firm conclusions. The summary and conclusions are anything but summary and conclusions. It lingers over 2 pages. Yet, from it, one can extract the two points of #9; diluted and tentative, as follows, quote:

a) The coincidence of Maunder's "prolonged solar minimum" with the coldest excursion of the "Little Ice Age" has been noted by many who have looked at the possible relations between the sun and terrestrial climate.

b) …the existence of at least two other major changes in solar character in the last millennium: a period of prolonged solar quiet like the Maunder Minimum between 1460 and 1550 (which I have called the Spörer Minimum) and a "prolonged sunspots maximum" between about 1100 and 1250.

If the prolonged maximum of the 12th and 13th centuries and the prolonged minima of the16th and 17th centuries are extrema of a cycle of solar change, the cycle has a full period of roughly 1000 years. If this change is periodic, we can speculate that the sun may now be progressing toward a grand maximum which might be reached in the 22nd or 23rd centuries…

The preceding text is the best we could extract from Dr. Eddy's "summary and conclusions." The following table is a direct translation of Eddy's message:

Roman Warm Age	Dark Age	Medieval Warm Age	Little Ice Age	Current Little Warm Age	Future Little Ice Age
+3^{0}C	-2^{0}C	+2.5^{0}C	-2^{0}C	+2^{0}C	-4^{0}C*
250 BC	250 AD	750 AD	1250 AD	1750 AD	2250 AD	2750 AD

I—— 1000 years — — | — — 1000 years — — | — — 1000 years —— I

* - 4^{0}C : The progressive slide into the next maxi-ice age.

Fig 37 . The Schematic Summary of 3 Eddy Cycles, covered in Part One.

* * * * *

The beauty of accumulating good cause-to-effect correlations between solar activity and mini-climate cycles is the development of good predictions. So far the main connections have been:

Point 4. 1645-1700. Maunder Minimum → Deepest Little Ice Age.

Point 5. 1700-1795. High Sunspot number → 18th Century Warming, Glacier Bay.

Point 6. 1795-1825. Dalton Minimum → Secondary Cooling now cyclical.
        1885-1925. Industrial Minimum → Slight Cooling (new name proposal).

Point 9. 1930-2000. Grand Solar Maximum → Strongest Warming of this mini-cycle.

The first to our knowledge to perceive a return to the Dalton event was East European Solar cycles mathematician Milivoje A. Vukcevic[21] who developed the following equation: fig.38   from Google 1-20-2020

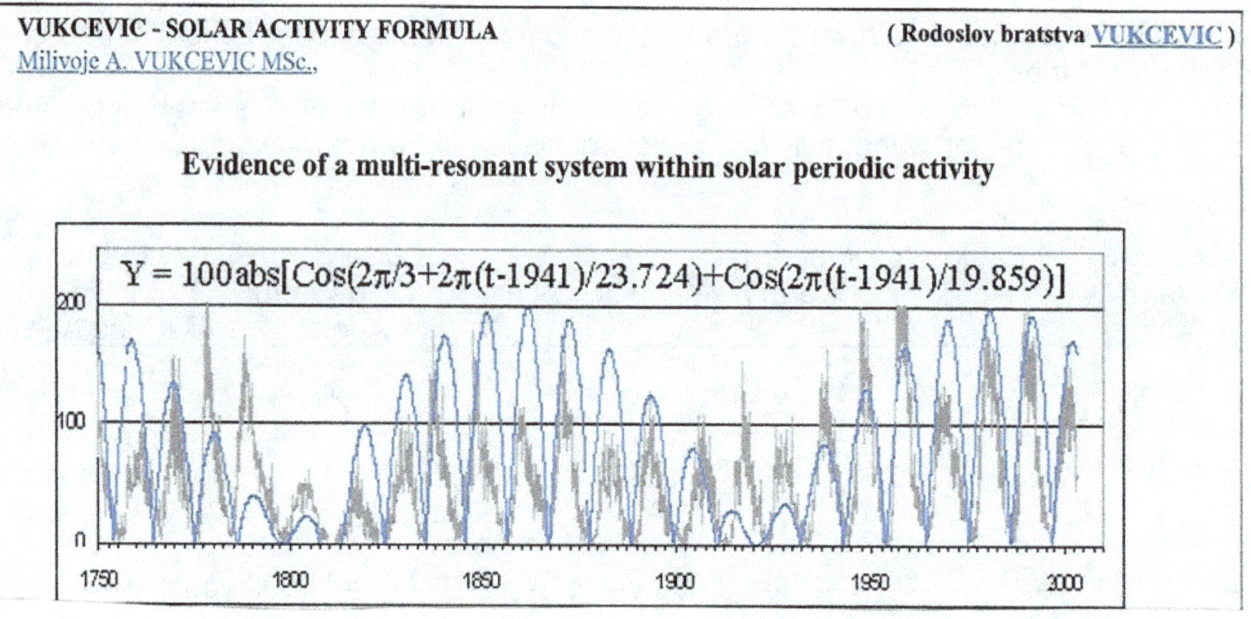

Fig 38. The blue curve is the application of the equation on the spot cycles (black)

In his 2011 book "Cold Sun"[23] John L. Casey reproduces the resulting prediction by Vukcevic of the second Dalton Cooling. Fig 39.

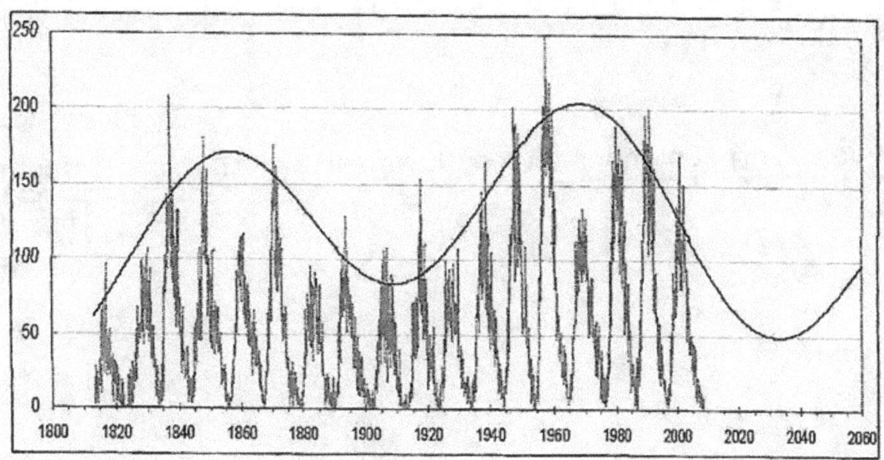

Figure 5-7. The 206-year Bi-Centennial Cycle over the last 200 years. Vukcevic's chart of the past cycles overlaid with his formula-derived curve for the 206-year cycle shows the coming solar hibernation low point at the far right. To the far left, around 1815, is the low point of the last solar hibernation during the Dalton Minimum. As of March 2011, we are headed down again toward the bottom in the 2030s *right on schedule*, as predicted. With it comes the cold. Source: Milivoje Vukcevic.

Dr. Habibulo Abdussamatov, Russian Academie of Sciences, even predicted the peak sunspots counts: Cycle 24: 70-75. Cycle 25: 50. Cycle 26: below 50

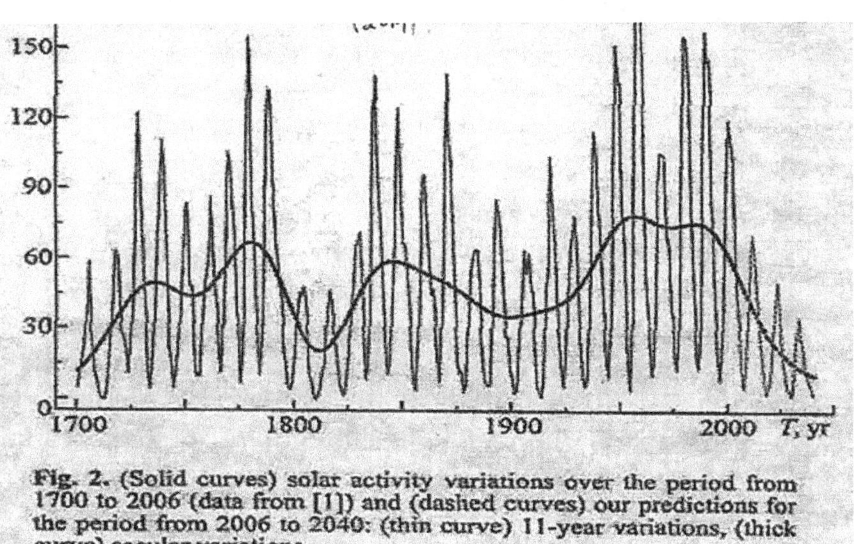

Fig 40.   The last 3 peaks are in dotted line predicting steep decline. This 2006 prediction by Abdussamatov was relayed from Casey.[23]

Let us not leap forward from those 2005-2010 predictions based on the "206-year Bi-Centennial cycle" shown in Fig. 39, before checking with recent data.
One of the most recent, well accepted, references comes from the Uccle Observatory in Brussels, Belgium[24]

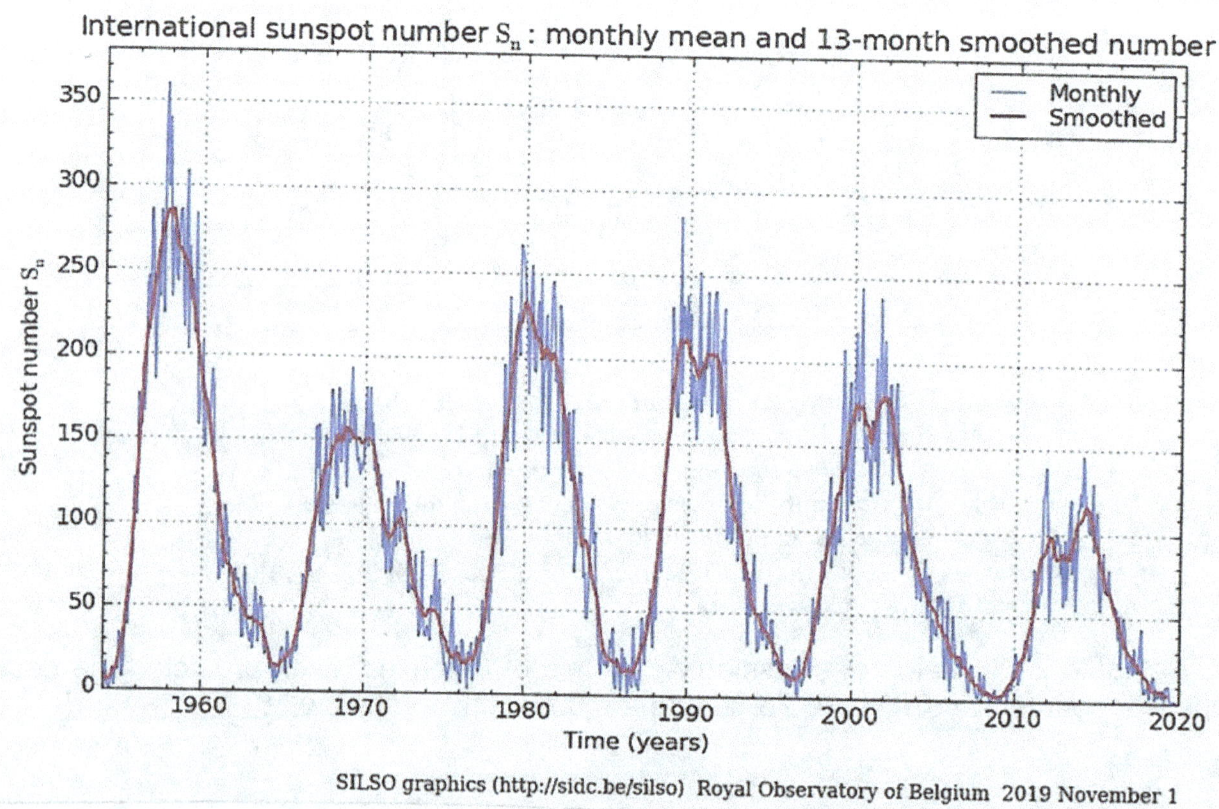

Fig 41. Dated Nov.1, 2019, this sunspot diagram shows several points discussed hereunder

1. The Abdussamatov prediction for cycle 24 peaking in 2014 was: 70-75. The reality was 120-130, still way down by almost half from cycle 23 peaking in 2000-2002.

2. The 2009 and 2019 bottoms are much lower than previous bottoms.

3. The down slope of cycle 23 from 2002 through 2009 is clearly longer than others and reminiscent of the lost cycle of the first Dalton Minimum from 1788 through1799. This is now becoming a signature of the onset of Dalton coolings.

These authors including the following diagram by Vahrenholt & Lüning[25] thus seem to be right in predicting the 2020 to 2045 cooling.

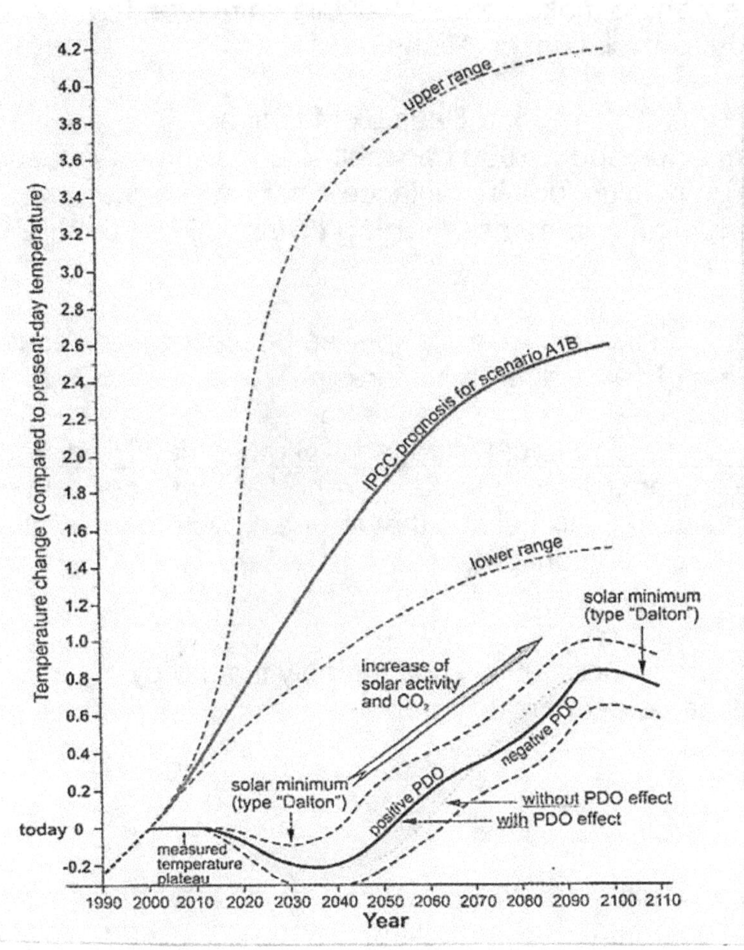

Fig. 42. The German 2012-2014 prognosis of temperatures in the 21st Century. It still gives CO_2 some action which we deny. PDO is for Pacific Decadal Oscillation. It also predicts the next Dalton to start in 2100 which seems to be 100 years too soon. The bottom at 2035 of the current second Dalton agrees with the other authors. From "The Neglected Sun". F. Vahrenholt and S. Lüning.[25]
The IPCC predictions, based only on CO_2, are clearly exaggerated.

All this seems to clash heavily with all media reports that average temperatures in 2019 were second highest ever, following a series of records since 2008.

To try and reconcile both sets of data from Madrid and from the preceding authors, let us sit back and observe the next decade. I believe that by 2025-28, the IPCC will have to adjust, excuse, find a way out to explain the cooling.

However, we tend to want to make sense of all this immediately.
Here are a few uncontroversial observations:

1. The ALBEDO Effect.   This is not a Deus ex Machina!
The evident warming since about 1950 has melted an enormous mass of ice around the North Pole. When dark blue ocean replaces bright white ice, the increase in **solar radiation retention** has an enormous **warming effect**.   Even the IPCC agrees!

Result?
The Dalton cooling is being delayed by this self-accelerating Albedo warming by, I estimate, 5 to 10 years. I call it climate hysteresis.

2. The Onset of the Dalton Sunspot Decrease is Slower than Predicted.
Compare the Brussels diagrams with the ones reported by Casey and observe that cycle 24 is higher than predicted. I still believe that the next cycle peak around 2025 will show a maximum of the order of 50 sunspots.

3. The Tendency to Reinforce.
The story of temperatures since 2000 is unavoidably tainted by the strong desire to prove a point. Climate change catastrophism at work!

 In conclusion of Point 11, we are facing a limited cooling of the earth's climate called the second Dalton Minimum. However the timing of the onset is probably delayed by some five to ten years by climate hysteresis. Later in the century, at least by 2045-2050, warming will resume following surging sunspot activity, clear through 2100.

* * * * *

<u>**12. 2014-2019. How Sunspots Impact Climate. Svensmark, Luyckx.**</u>

By now, it is obvious in Google and Wikipedia that a growing number of scientists at NASA and elsewhere believe or, at least, suspect that sunspots impact on our climate. A good example of today's tongue-in-cheek allusions to the connection between sunspots and climate is given in fig. 43 below.

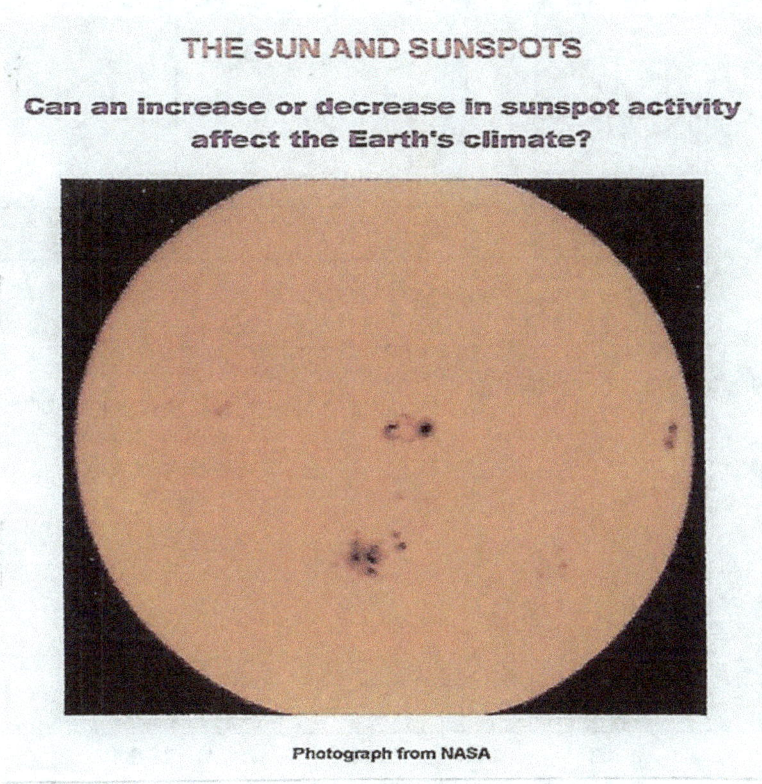

Fig.43. A 2013 NASA picture of sunspots. Please notice the elliptical shape of the spot on the right, ready to move to the other side. This is what Galileo observed with his new 33x telescope in the fall of 1609.

So, sunspot frequency affects climate. But what is the mechanism?

The first approach is evident. Let's look at the sun's energy output without our atmosphere's interference. Since about 1980, at least one satellite has been solely dedicated at measuring the solar output with great precision. This output is remarkably constant at 1,366 plus or minus 1 or 2 Watts per square meter. Very disappointing is the non-response to sunspot peaks and valleys. The difference of plus minus 0.1% from sunspot cycles is far too small to explain any climate change from sunlight.

Two diagrams from Jack Eddy's 2009 book[26] (p.110) on the sun show it well:

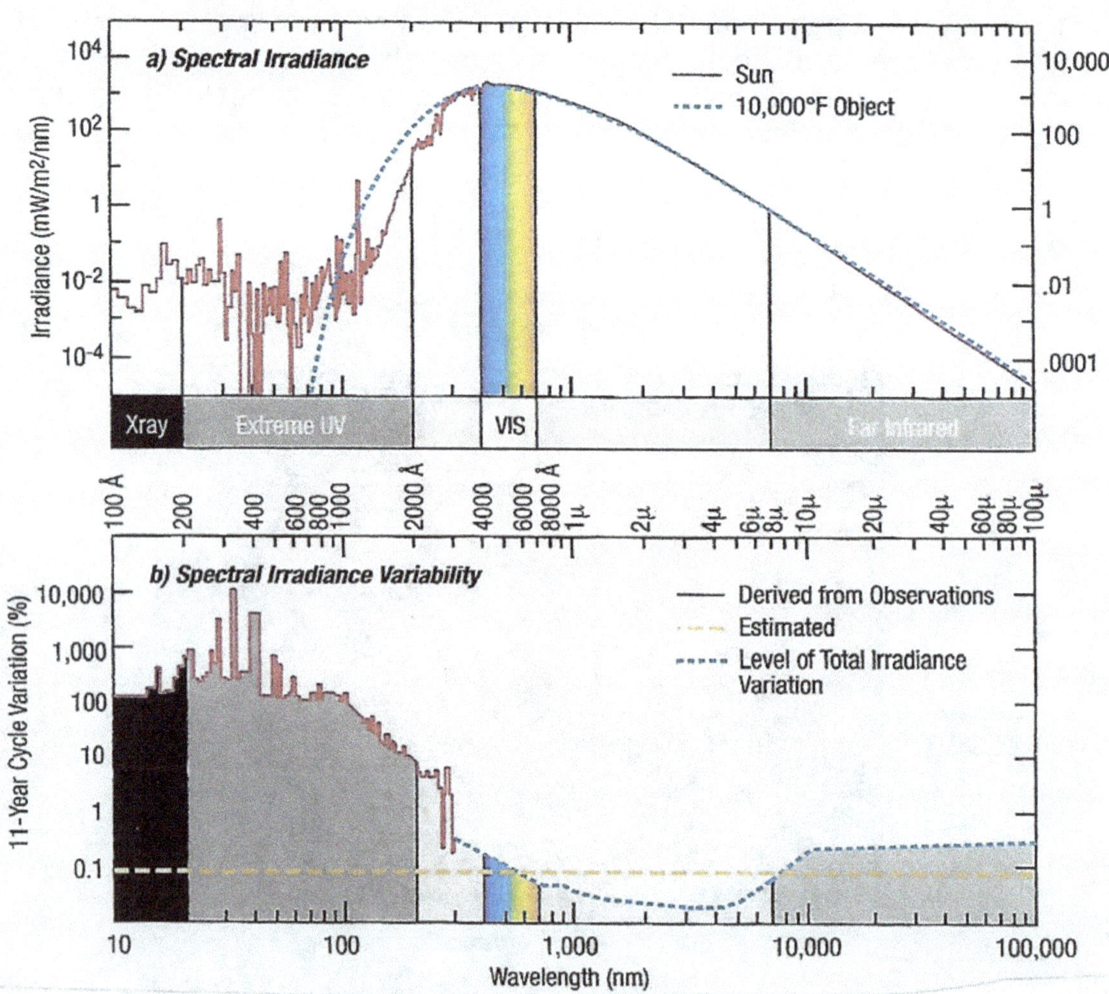

Fig 44. The top diagram shows the sun's radiating energy as a function of the wave length. It is clear that most of it is between 1000Angstroms and 100 Mu.

The bottom diagram shows the variation caused by sunspot cycles. It is clear that all that variation occurs in the short wave length range where almost no energy is present. Where the energy is, only 0.1% variation shows from the sunspot cycles.

The following figure also from Eddy's book[26] shows the effect on watts/m^2

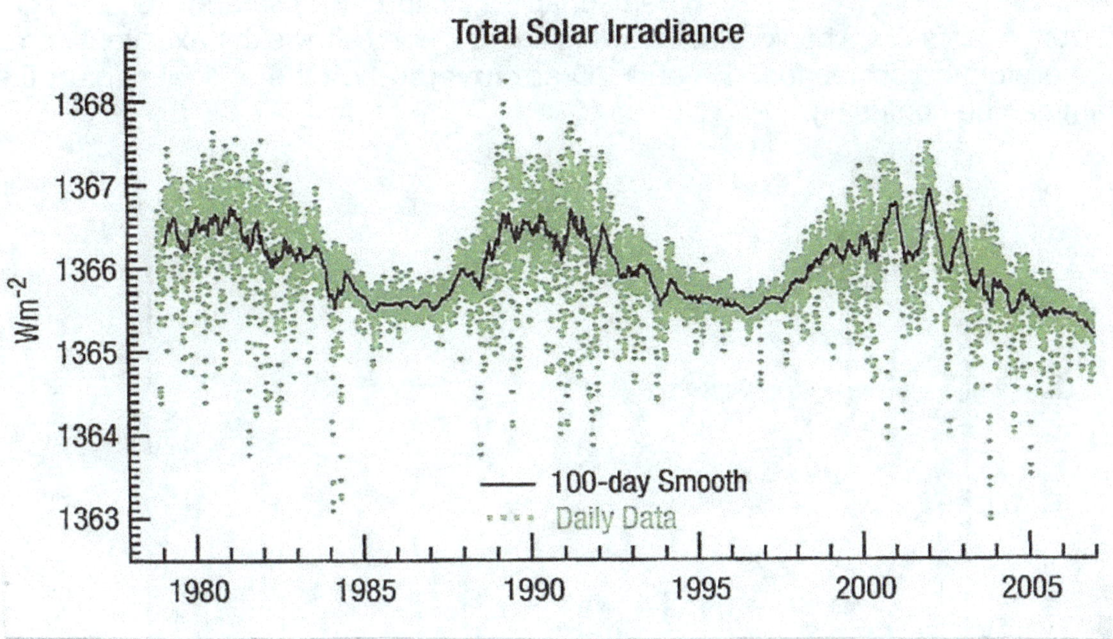

Fig 45. Total Solar Irradiance – except the Magnetic Vector introduced by the author – clearly varying with sunspots cycles but, only by 0.1%, totally insufficient to explain climate change.

So, that sort of energy cannot account at all for the global warming caused by sunspots.

What is it then?   Enters **Svensmark!**

Fig 46

Dr. Henrik Svensmark, Danish, born 1958.

A Physicist and Professor at the Danish National Space Institute and Solar Researcher.

Svensmark studied extensively the Beryllium 10 contents in paleo-ice cores from Greenland and Antarctica and Carbon 14 in ancient dated tree rings. Among others, he pointed out the clear relation between sunspot activity and these two variables:

When sunspot count is high, ^{10}Be and ^{14}C are low.
When sunspots disappear, ^{10}Be and ^{14}C are high.

Again from Eddy's 2009 book[26] the following figure clearly shows the excellent correlation but be careful: the scales for ^{10}Be and ^{14}C are **inversed** on the ordinate so, at first, the diagram can be confusing.

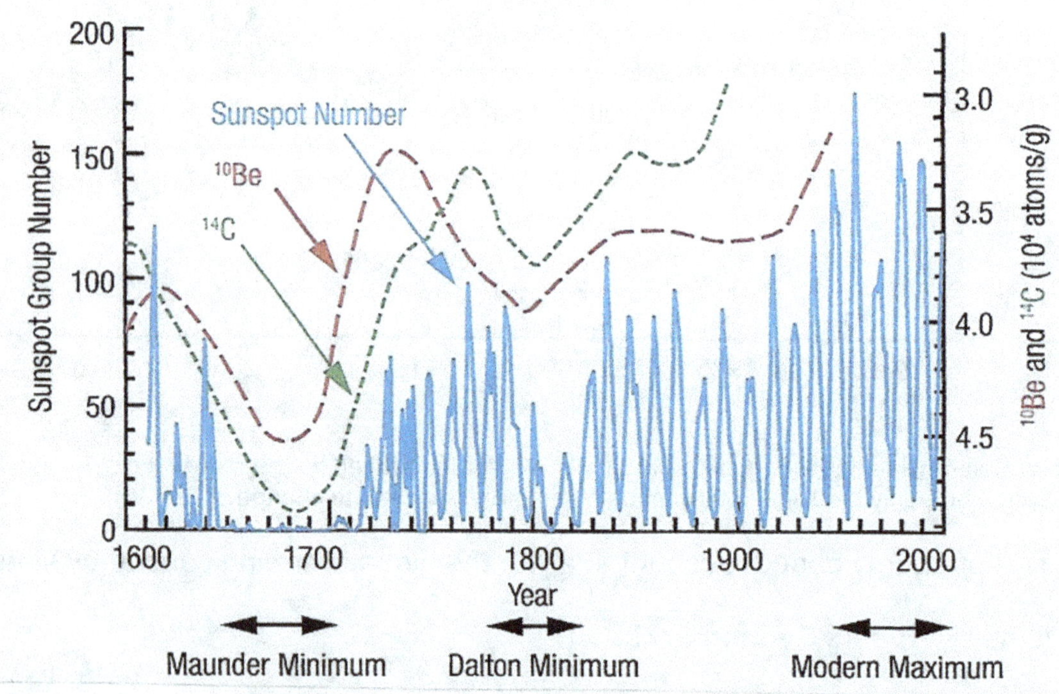

Fig 47. The outstanding correlation between sunspot numbers and ^{10}Be in Greenland and Antarctic ice cores, and ^{14}C in dated tree rings. Be careful to inverse the ordinate for Berylium and Carbon. What looks like low green and red peaks are actually high peaks, etc..

Svensmark's proposal is as follows for the mechanism of cooling and heating:

^{10}Be and ^{14}C are the result of high energy cosmic rays from our galaxy, invading our upper stratosphere and colliding with oxygen and nitrogen atoms. These two isotopes tend to nucleate cloud formation on their way down to earth. When sunspots are high, the magnetic field sent by the sun stops many cosmic rays and diverts them. So, low cloud formation. When sunspots disappear, cosmic rays have a free way in and nucleate a lot of clouds, blocking the sun's rays from reaching the earth's surface.

The Svensmark theory was not universally accepted at first in the 1990s but the Danish team has made progress.

In his 2019 book,[6] "***The Solar Magnetic Cause of Climate Changes and Origin of the Ice Ages***"Dr. Don J. Easterbrook schematizes the Svensmark process of cloud formation as follows.

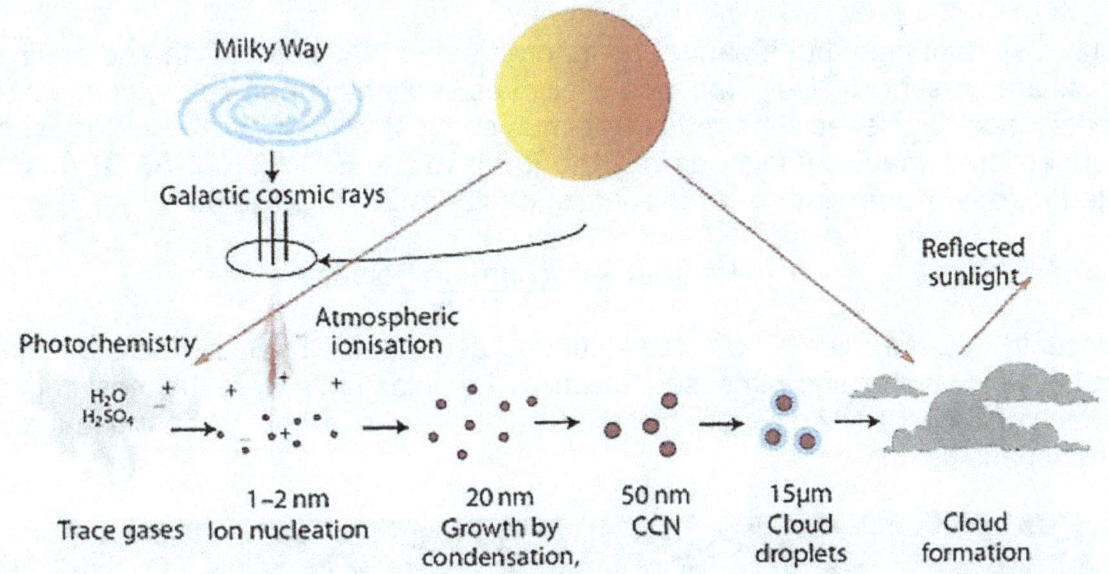

In summary, the link is: (a) a more active Sun, (b) stronger solar wind, (c) fewer cosmic rays, (d) less atmospheric ionisation, (e) less nucleation and slower growth, (f) fewer CCN, (g) clouds with less droplets, (h) less reflectivity, (i) less reflection of sunlight and a warmer Earth.

. Diagrammatic summary of the Svensmark theory of climate change.

Critics of the Svensmark hypothesis claimed that particle clusters produced by cosmic rays measured only a few nanometers across, less than the needed 50 nm aerosols typically need to produce cloud condensation nuclei. Further experiments by Svensmark and others published in 2013 showed that aerosols with diameters larger than 50 nm are produced by ultraviolet light (from trace amounts of ozone, sulfur dioxide, and water vapor), large enough to serve as cloud condensation nuclei.

Fig 48. Dr. Easterbrook's rendering of the Svensmark process from ion nucleation to cloud formation and the recent progress made by the Danish Solar Research Academy to uplift their theory.

Dr. Easterbrook and others endorse Dr. Svensmark's views in part because it is, as of today, the only existing mechanism available to account for sunspot's warming action. I join in because of the impeccable logics of the theory but I believe that this clouding action or non-action is less than 10% of the total process of warming by sunspots.

How do Sunspots Warm the Earth?   **The Luyckx Mechanism.**

It is only visible at night but it cannot be ignored at high latitudes north and south when sunspots are present. It flies in the face of all of us and it has been solidly connected with sunspots since the Benedictine monk Adelmus suggested in 807AD! So, for everyone, the connection is matter of fact has **but the jump to ENERGY TRANSFER has**, to my knowledge, **never been made until my first essay in 2014**, **page 43**

The Polar Lights. (Aurorae Polaris)

Just recently – December 2019 – the returning astronaut, Christina Koch, emphasized that out of all the sights she witnessed in space, the polar lights were the most impressive and beautiful. The following picture by NASA gives an idea of what this astronaut admired from her space station.

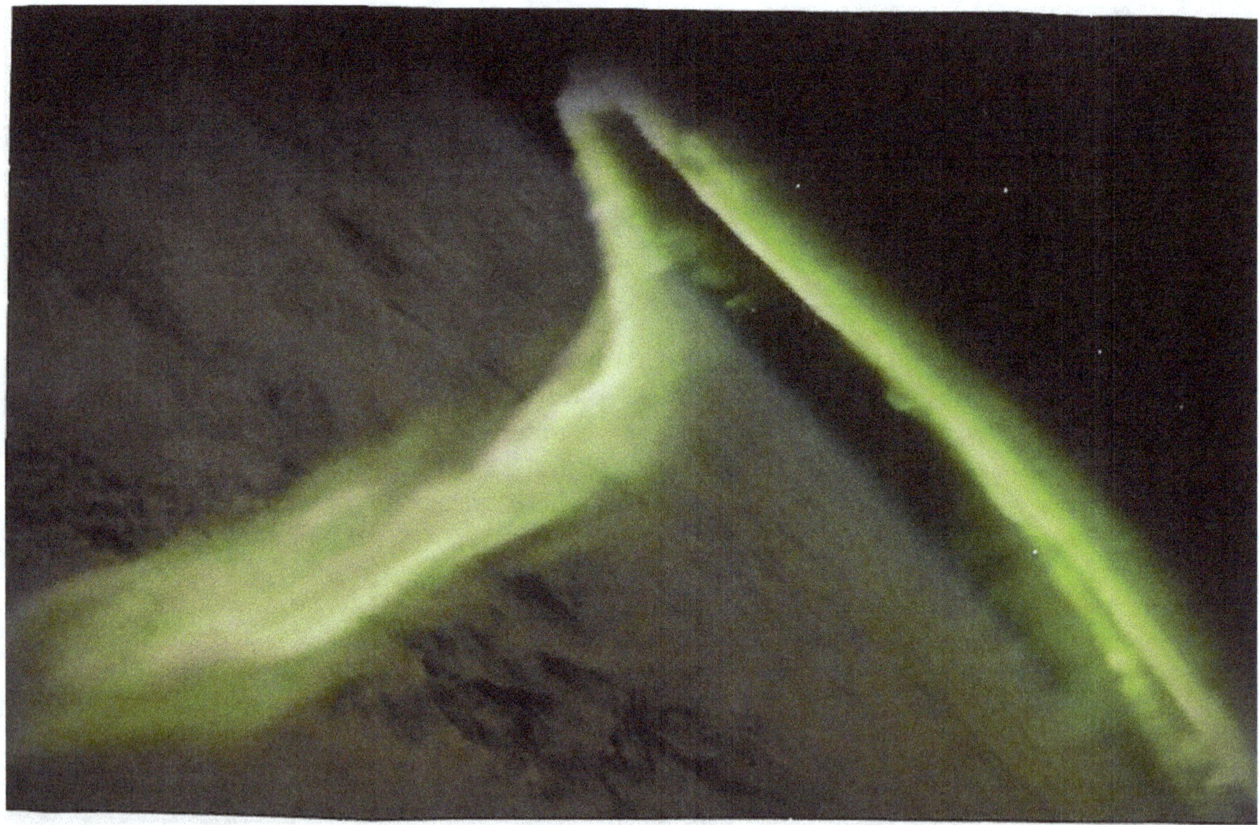

Fig 49. From space, a NASA photograph of a Polar light illuminating the Earth surface below. Notice the fuzzy top and the clear cut bottom of this luminous display of magnetic energy.

A. First: <u>Polar Lights are caused by sunspots.</u> No polar lights during the Maunder Minimum

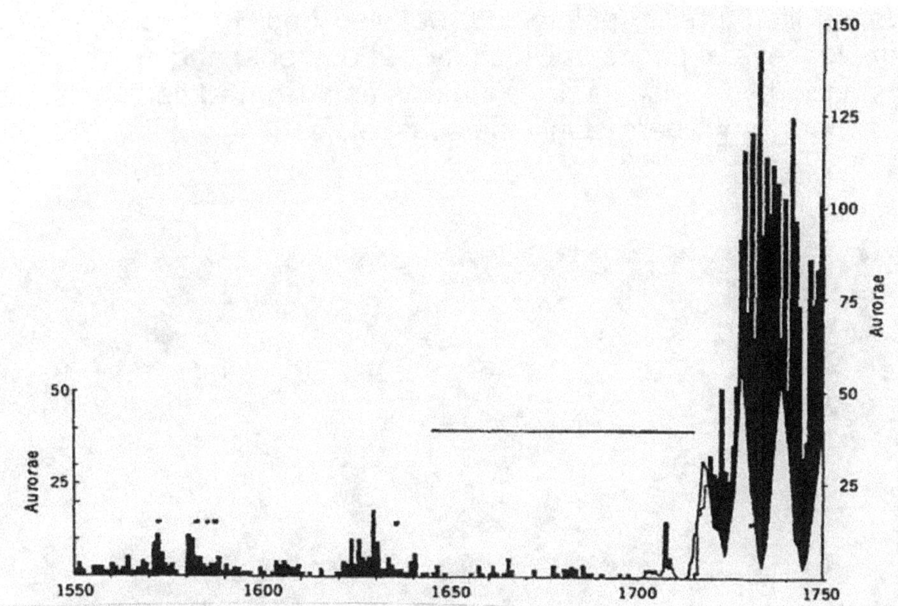

Fig 50. It is clear that during the 1645-1700 Maunder Sunspot Minimum, aurorae all but disappeared and returned together with the sunspots after 1700. The Maunder Minimum, by Dr. Jack Eddy. [2] 1976.

B. Second: <u>The Warming Comes from the Poles.</u>

For at least two decades now, the media have observed and ascertained by countless ice melting reports that **our warming "comes from the Poles".** CO_2 can certainly not be involved here. Maybe CH_4 is the culprit, reputed to be 6 to 25 times worst than CO_2 – depending on articles - because of permafrost melting. However, as we have seen in point **9** with the Mauna Loa station, our atmosphere mixes so fast that the CH_4 dilutes within weeks to world averages, down to the middle Parts per Billion range, 4 orders of magnitude lower than the 400ppm CO_2 concentration. **So greenhouse gasses are excluded from this polar warming mechanism.**

C. Third: <u>Warming at the Poles Has to Be Magnetic.</u>

It is unquestionable that the Polar Lights are concentrated on the Poles by the earth's magnetic field. Something coming in from the sun is strongly co-axed on our magnetic poles by the earth's field. It just happens – has always and will always happen – that the geographic poles – 90° latitude – are very close to the magnetic poles. The distance varies with time but stays within 10 degrees latitude from the geographic poles. Sometimes, the magnetic poles switch polarity. The last one was about 750,000 years ago. However, the axis of magnetism stays close to the axis of rotation of the earth. They are interrelated.
So, whatever causes the warming at the poles has to be of magnetic nature.

D. Fourth: <u>The supposed "**Magnetic Field** of the Sun" Probably not Unified.</u>

The Solar Science literature is replete with detailed depictions of the "sun's magnetic field". For example, several pages in Dr. Eddy's 2009 book, and its effect among others on incoming galactic rays. Just suppose it exists as a **unified** field, it cannot deliver, as such, a Watt of energy anywhere. Only the sunspots can.

A magnetic picture or magnetogram of the solar photosphere made (from 93 million miles away!) on October 28, 2003 showing regions of strong magnetic polarity. White portrays what is conventionally called *positive* (or *north*) polarity, black *negative* (or *south*) polarity. As we see here, magnetic regions (which correspond to regions of sunspots and other manifestations of concentrated solar activity) are made up of adjacent parts of opposing magnetic polarity which are confined within two distinct belts of solar latitude.

Fig 51 . The polarity of sunspots (North=White, South=Black) shows the Magnetic complexity of the sun's surface and strongly advocates against a unified solar field. From Dr. Eddy's 2009 book. [26]

Before proceeding further, let us explore who has tried to quantify the magnetic strength of these darker spots on the face of the sun. Starting about in 1998, at the National Solar Observatory in Tucson, AZ, two NASA scientists Matthew Penn and William Livingstone used the Peter Zeeman effect on the spectral lines of iron to estimate in B Gauss the magnetic field near the center of sunspots. Iron, by the way, is the result of countless impacts of large and small meteorites over the 4.5 billion years of the sun's life.

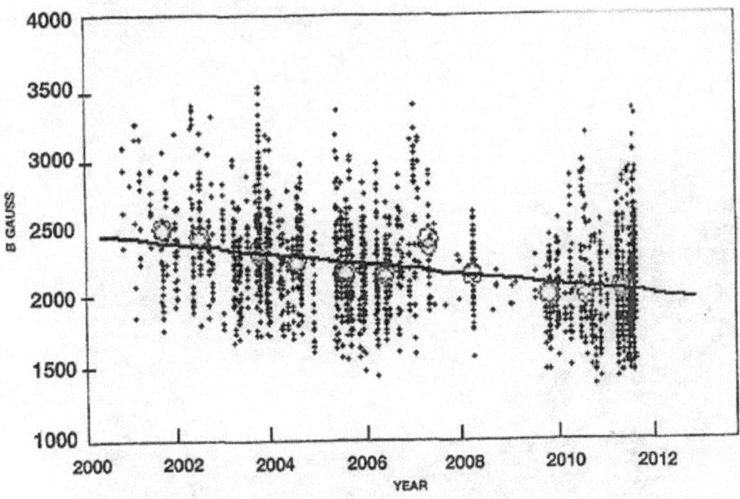

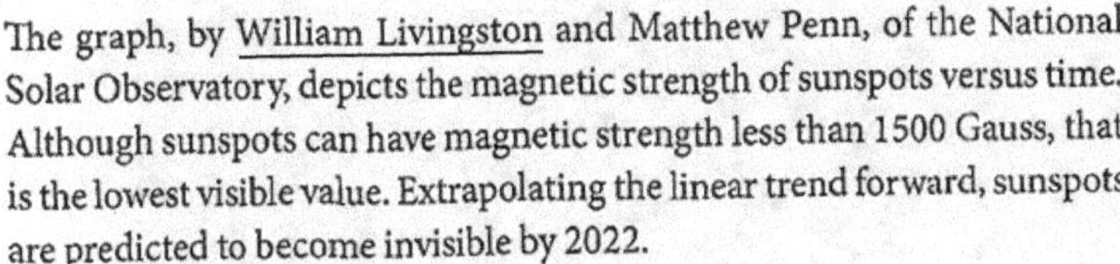

The graph, by William Livingston and Matthew Penn, of the National Solar Observatory, depicts the magnetic strength of sunspots versus time. Although sunspots can have magnetic strength less than 1500 Gauss, that is the lowest visible value. Extrapolating the linear trend forward, sunspots are predicted to become invisible by 2022.

Fig 52. From the book "Don't Sell Your Coat" by Harold Ambler.[17] 2011. Please compare these enormous Gauss levels to the less than One Gauss for the magnetic field of our planet.

Ten years ago, there was a school of thinking that the down slope of the magnetic measures average as seen on the diagram predicted a complete extinction of the sunspots by 2016. This evidently did not occur. However peak #25 is predicted to be the lower than the 125-150 counts of peak #24 in 2014. It should only reach 50 or less by 2026.

The absence of sunspots between cycle 24 and 25 has lasted almost a year so far in 2020.

The net overall result of this NASA study is to introduce a new sunspot factor: Their strength. So, not only will sunspots of cycle 25 be few but they will also be week, which further lowers the magnetic energy sent to earth.

Norðurljós (aurora borealis)

Fig 53.   Omega shaped northern light. Picture purchased at Keflavik airport.

This very picture of polar lights points to streams of impinging matter reacting luminously when encountering our upper stratosphere. At first it is a fuzzy display. Then it amplifies further down, and reaches a critical frequency of atmospheric particles below which the contained energy cannot survive. That's the bright bottom of most polar lights. The bottom is the brightest and cut short.

This description clearly indicates <u>the morphing of magnetic energy carried by the incoming stream of matter to light, the polar light</u> upon contact with atoms of oxygen and nitrogen.

This morphing of magnetic energy is not just to light but also to invisible heat, my central hypothesis for mini-climate change, as highlighted on the front cover.

Note: A PhD thesis could be based on the Zeeman measurement of magnetism of these polar lights as well as their light spectrum.

We have entered here virgin scientific territory that I am not equipped to handle even to suggest an approach to prove or disprove the hypothesis.
There are problems: How does that heat, if it exists, transfer to the polar surface from such high altitude? Infrared radiation? The colors of the polar lights are only infrequently red. Would it be just those which warm our poles? On the other hand, for energy transfer to earth, it does not matter whether these polar lights are visible or not. It is a 24-hour process that continues even when the sun shines over the poles.

F. Sixth: Sunspots Eject Huge Invisible Magneto-charged Streams of Matter.

So many ejections from the sun are highly visible. Explosions, flares, the corona, etc.. How come those emitted by sunspots are invisible? First of all, **it is only the magnetic vector** of those violent emissions from explosions that create havoc on our electric grid and airplane compasses. All of the highly energetic and luminous matter ejected **returns to the sun's surface** if it is not magnetically charged. **No matter escapes except magnetic.** The permanent ejection of matter from the sun is very improbable because of its enormous gravity. All those big luminous streams of ejecta clearly return to the sun because of gravity. The energy of sunspot matter, however, is entirely magnetic but invisible. What is the force that permits such matter to escape the solar gravitational field? The 2500 gauss magnetic polarity of the matter ejected is surrounded by a "cylinder", extending fairly deep into the sunspot, of the same polarity. The repulsion force is so great and extended into space that the matter keeps escaping the sun's gravity.

The strongest sunspots exhibit a spiraling vortex (fig 11) boosting the speed of the jettisoned material to space. The whole violent ejection process extracts Terawatts of energy very locally, resulting in 1,000 to 1500° C cooling, darkening the spot to apparent black.

From many accounts of the time lapsed between sunspots appearances and polar lights the consensus is: it takes 10 to 15 hours for that stream of matter to reach our upper stratosphere. This means of the order of 1% of the speed of light or 3,000km per second on average. My suspicion is that the speed of ejection is much higher at exit from the sunspot and keeps decreasing into space because of the charging balance between magnetic repulsion and the sun's gravity.

G. Seventh: <u>How many Watts/m^2 Reach the Poles From Sunspots?</u>`

In the absence of any direct measurements yet of polar lights, the best available evidence is the Maunder Minimum. Indeed, sunspots all but disappeared for 55 years, 1645 to 1700. The effect was the coldest of 3 little Ice Ages and an estimated 2°C drop. We don't have direct sunspot count from the previous two, the Wolf, 1300-1375 and the Sporer, 1450 to 1540, because the telescope had not been invented. The Dalton cooling does not deliver enough temperature differences. We also know that, before the Maunder Minimum, we were still, probably 0.3 to 0.5°C below neutral because of anemic sunspots counts by early telescopes from 1609 to 1645. This is part of the entire 1250 to 1750 mid-millennium cooling period of the Eddy cycle.

Back to the Maunder Minimum! Suppose the spot disappearance had lasted 100 years or 200 years or "forever". Would the cooling have been worse? Or would it have stabilized, after 50 to 70 years to a low temperature "plateau"? In fact, we don't know. Implicitly assuming that plateau, however, appears illogical and far less probable than expecting a continuing cooling from sunspots disappearing longer.

In his 2019 book, Dr. Easterbrook is the first to assign sunspots disappearance as cause of the maxi ice ages as well as of the little ones. The Maunder Minima would have to extend to hundreds and thousands of years. If this is correct, it has a major implication on the energy received by the poles from sunspots. The major part of the energy remains that of light at 1,366 watts/m^2 but the magnetic energy fraction has to be much higher than the 0.1% difference in light energy.

Let us continue with observations and logical deductions! The coldest ice age periods in ice cores are -8°C from neutral. We have to add +5°C for the hottest interglacials.

If we have to account for all that 13°C variation in terms of energy emitted or omitted by sunspots, we arrive at some serious watt figures per m^2. As the sunlight part of the energy received remains essentially constant at 1,366 W/m^2, that fraction of the total diminishes to explain the 13°C difference.

I suggest to peg at one half percent of the total solar energy, the loss or gain of 1°C on earth.

The difference between the two extremes or 13°C would correspond to 6.5% of 1,366 W/m^2 or 88.8 W/m^2. So, at the bottom of ice ages, the solar energy would be only sunlight at 1,366 W/m^2 while at the early top of interglacial times, the total solar energy would be 1,454.8 W/m^2, of which 88.8 or 6.5% from sunspots.

As we are, early in the 21st Century, just past the middle of our Little Warm Age, we can estimate being some 1.5°C over neutral or 9.5°C over glacial bottom. We are thus receiving, on average, 88.8 x 9.5/13 = 64.9 Watts/m^2 from sunspots.

Now, this 65 W/m^2 "average" is variable in time with sunspots cycles and space, the poles receiving most of it. Right now in 2020, near the zero line between cycles 24 and 25, the earth receives only the 1366 W/m^2 sunlight fraction. Barely any spot is visible and polar lights are rare. On the other hand, in 1958, at the highest sunspot peak ever recorded, the magnetic energy received on the poles may have topped 200 W/m^2. Sounds enormous but the warming effect is not immediately visible. There is considerable climate hysteresis, delay and effect of ocean currents, jet stream, etc.. The same is true today. It will take one half to two thirds of this decade to observe unquestionable cooling, at or after the low 50 sunspots peak of 2025. This entire subject is developed in detail in Part Four.

H. Eight: The Signature of Temperatures over the last 400,000 Years. The Maxi-cycles.

Along with Dr. Easterbrook, we have now accounted for the enormous 13°C in terms of energy and effects. Supposing the 6.5% difference in total irradiation is about correct, this difference cannot be assigned only to the outer 20-30% volume of the sun. A sunspot absence for centuries or millennia has to result from a corresponding decrease in hydrogen fusion energy at the core. All eight glaciations over the past 800,000 years offer the same temperature signature. That signature is sending us a strong message on how the sun's radiation evolves in our Pleistocene epoch. Because each glaciation repeats the same pattern, the message is reinforced. The best diagram I could find to show this pattern is from the 2011 book: "Don't sell your Coat" by Harold Ambler.[17] The data were obtained so far back are from a universally accepted proxy: the ratio of heavy isotope ^{18}O over the more abundant, normal ^{16}O. That ratio is consistently tied to the temperature at time of burial of ice in the miles thick Antarctic ice sheet.

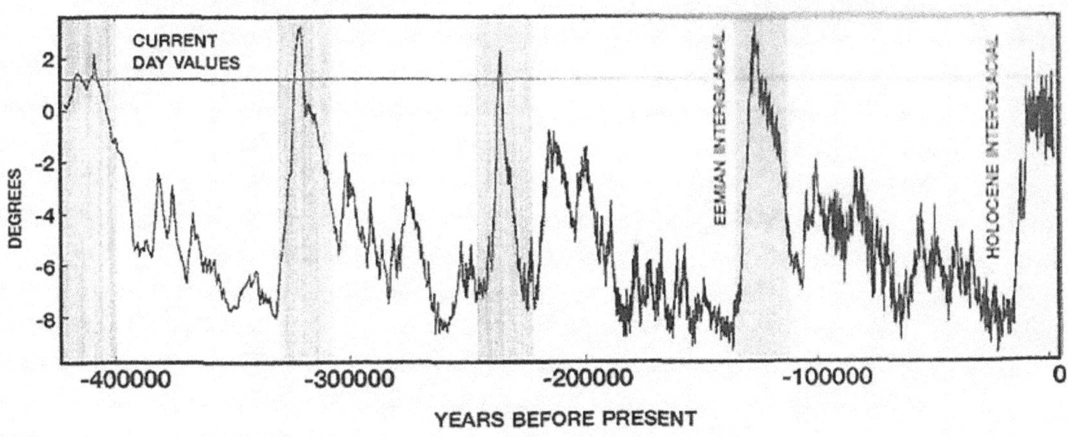

Fig 54. The characteristic temperature of our planet over the last 400,000 years. Notice the narrowness of the interglacials as compared to the length of the glaciations. The shaded areas are wider than the actual 5 t0 8,000 year peaks of each interglacial.

My interpretation of the temperature signature is as follows:
A brief period of high activity exhausts the sun's capacity to continue at that pace. The interglacial is very short. Somehow, the hydrogen immediately available for fusion is reduced progressively. The long glaciation begins with considerable ups and downs in which Singer and Avery thought having detected a 1500 year cycle in their 2007 book entitled "Unstoppable Global Warming. Every 1500 Years".[27] It is, however, unescapable that the general trend during each glaciation is downwards. **The lowest temperature of each glaciation is at their very end.** Then a very sudden surge occurs each time going straight to the peak of the interglacial.

Somehow, the sun slowly but surely contracts over 70 to 90,000 years with ups and downs and reaches at the end of the glaciation a trigger point of compression that suddenly reactivates a flurry of new hydrogen fusion, launching the next interglacial. The $12°$ to $16°C$ surge occurs in a few hundreds to a couple of thousands of years.

This signature negates any possibility of earth centered models to be valid, such as Milankovic's axis rotation. The surge of energy required to climb $13°C$ in a thousand years can only come from the **SUN**.

* * * * * * * * * * * *

According to classic philosophy, the sun is perfect, unblemished and constant. Yet dark spots on its surface were observed and recorded since antiquity. The invention of the telescope in 1608 allowed detailed counting of sunspots and triggered a priority contest promoting assiduous recording. Mankind is lucky that this happened 36 years before these spots all but disappeared for 55 years. Called the Maunder Minimum 3 centuries later, this quietness of the sun resulted in the coldest of 3 mid-millennium little ice ages, the Wolf 1300-1380, the Sporer 1450-1540 and the Maunder 1645-1715.

Sunspot activity surged up from 1700 to 1795, causing substantial warming as demonstrated at Glacier Bay. A secondary decrease of activity, later called the Dalton Minimum, cooled the earth from 1795 to 1825. Except for a smaller depression from 1875 to 1925, which I name the Industrial Minimum, sunspot activity in eleven year Schwabe cycles kept increasing, peaking clearly from 1930 to 2001. This coincided with CO_2 rise in our air. We are now witnessing the onset of the second Dalton minimum from 2020 to 2045. It took centuries for someone, in 1976, Dr. Jack Eddy, to recognize that the 1645-1700 minimum caused major cooling and to organize these mini-climate events into a 1000 year cycle of warming and cooling called Eddy cycle.

Multiple evidence clearly shows that sunspot activity determines our mini-climate cycles but recent satellite measurements of total solar irradiation only show a 0.1% effect of sunspot cycles. So climatologists and solar physicists are looking elsewhere.

Dr. Hendrick Svensmark has proposed the galactic particles impinging our stratosphere as nuclei for cloud formation to be the variable cause of cooling. At sunspot peaks, the magnetic protection from the sun reduces the frequency of these particles and less clouds are forming, thus warming occurs.

I propose a more direct interaction which will be covered by the conclusion of PART TWO. I go further to suggest a mechanism, deeper in the sun, to account for maxi-ice ages.

****

<u>**Conclusions of PART II**</u>

In the absence of factual proof, it is essential to arrive at circumstantial evidence connecting sunspots directly to warming of the earth:

1 - With a 10 to 15 hours delay, sunspots generate our Polar Lights (Aurorae). Thus, there is ejection of material charged with energy from sunspots.

2 - The earth's magnetic field concentrates those streams of ions, protons and electrons on our Magnetic Poles. Thus that material is charged with magnetic energy.

3 - The display of light at the poles demonstrates, following the second principle of thermodynamics, a degradation of this magnetic energy to light and probably more.

4 - Multiple media evidence demonstrates that our warming originates at the Poles. Thus Polar Lights generate Heat.

The factual proof will come from detailed spectral and magnetic analysis of the Polar Lights from space satellites and from earth stations in polar regions.

I tried to quantify the magnetic energy added by the sunspots to the constant 1366 W/m^2 sunlight energy. Assuming 1°C resulting from ½% of sunlight, the grand average sunspot energy adds 65 W/m^2 which easily translates to well over 200 W/m^2 over the poles at peak sunspot activity. The transfer of that energy from the poles to the rest of the planet's surface takes time and is obscured by multiple cyclic ocean currents and other factors. It can be called climate hysteresis.

****

<u>**Last Minute News on Sunspots Observation. The New Maui Telescope.**</u>

I had no idea that a brand new – 4 meter primary mirror – solar telescope was being built on the Maui Island of Hawaii, called the Daniel K. Inouye Solar Telescope. It was built, from 2013 to 2020, by the National Science Foundation specifically to improve detailed sun surface observation and studies. The following picture was released December 4, 2020.

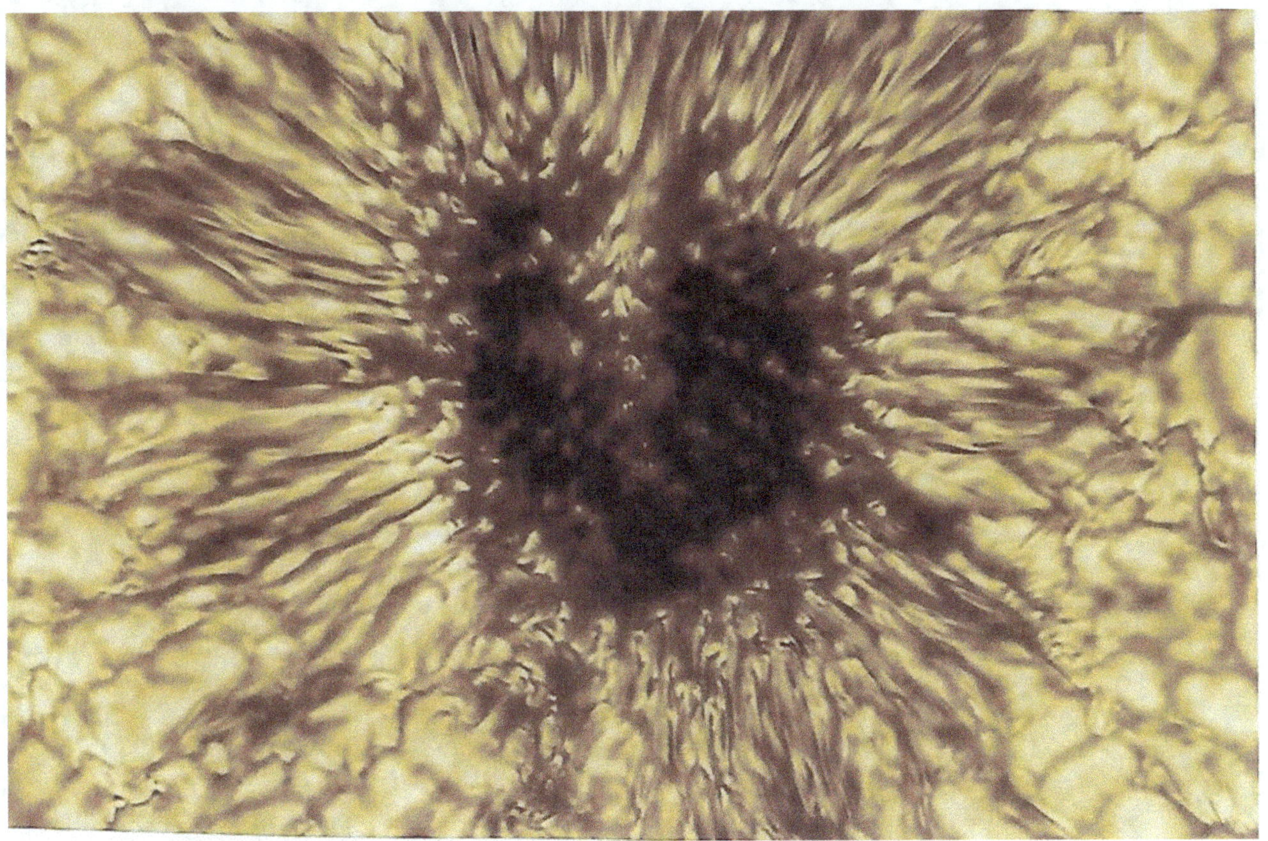

Fig 55. This first sunspot image by the Inouye Telescope dates back to Jan 28, 2020. It is a small spot as compared to the recent group of spots observed late November. Early December 2020 is marking the onset of Cycle # 25. See Google for details.

** ** ** ** **

PART THREE

The World's Belief on Climate Change: CO_2 + CH_4

The 33 year old Dogma ex-IPCC Cathedra:

Humans are Guilty of Climate.

This is a typical Medieval man and earth centered view of a sun centered phenomenon. It reminds us of Galileo against the Pope. The penalty to non believers is still excommunication. This time however, the accusation is "dangerous climato-skeptics". New York's United Nations replaces Rome's catholic Popes as the world's center of power on climate.

1.   <u>The Greenhouse Gas Effect of CO_2 and , lately, CH_4 on Climate.</u>

The solid and liquid surfaces of the earth receive over 99% of their heat from the radiation of the sun, mostly sunlight. At night, some of that heat radiates back to space as infrared radiation or, more simply, invisible heat. During cloudy nights, that back to space radiation is cut by over one half because the clouds send the infrared rays back to the surface. Everyone of us has experienced, particularly in the winter, this warm feeling on cloudy mornings as compared to frigid sensations after stary nights.

The greenhouse effect, however, is the invisible action of transparent components of the air, namely water vapor, H_2O; carbon dioxide, CO_2 and methane or natural gas, CH_4. The blocking effect is much smaller than from the clouds but real. The greenhouse Effect is

Proportional to Concentration

of these molecules in our atmosphere.

Water vapor is a very real greenhouse gas because its content in the air is in several percent. However, climatologists barely ever mention water vapor because it varies totally out of control from mankind.

H_2O vapor can be 1 in 10. CO_2 one molecule in 2500. CH_4 one in millions of air molecules.

These are the numbers greenhouse proponents keep hidden from the media.

To show how important the concentration of the greenhouse gas can be to its potential effect on climate, let us start with the extremely low, methane, CH_4 content of 100 to 400 parts per billion, ppb.

CH_4

The three main sources of methane escaping in the atmosphere are:
1. Natural gas leaks around poorly engineered ground piping.
2. Permafrost melting in Siberia, due to our warming climate.
3. Farm animals exhaling methane due to rumination for milk production.

I am not aware of the relative contributions of these three sources but the result in the air is so infinitesimal that it does not matter. In addition, our 21% oxygen content of the air makes methane unstable. It eventually oxidizes following the next reaction:

$$CH_4 + 2\ O_2 \rightarrow CO_2 + 2\ H_2O$$

To prevent the disregard of methane as a significant greenhouse gas, climate researchers are constantly upgrading their estimate of the greenhouse power of methane. Comparing that power to that of CO_2, they started with a factor of 6. Months later, it grew to 12, then in 2020 it rose to 18 and as of April 24, 2021 *the Economist* reported the factor to be 80.

The content of CH_4 is one thousand to 2500 times less than CO_2 in the air. So even at "80 times CO_2 greenhouse power" CH_4 cannot have more than $0.004°$ C effect on climate. This is of course totally negligible! The farmers should be congratulated for their milk production.

The recent paroxysm of attention and politics against beef consumption is ludicrous when compared to the numbers I demonstrated: less than 0.004°C. Only a little better numbers education could prevent at least some in the media from joining this stupid anti-beef bandwagon.

CO_2

Definitely the result of fossil fuel burning, the rise of carbon dioxide in our atmosphere since 1850 amounts to:

$$280 \text{ ppm increasing to } 420 \text{ ppm today}$$
equivalent to 0,028 % increasing to   0,042 %
equivalent to 1/ 3,571 increasing to   1/ 2,381

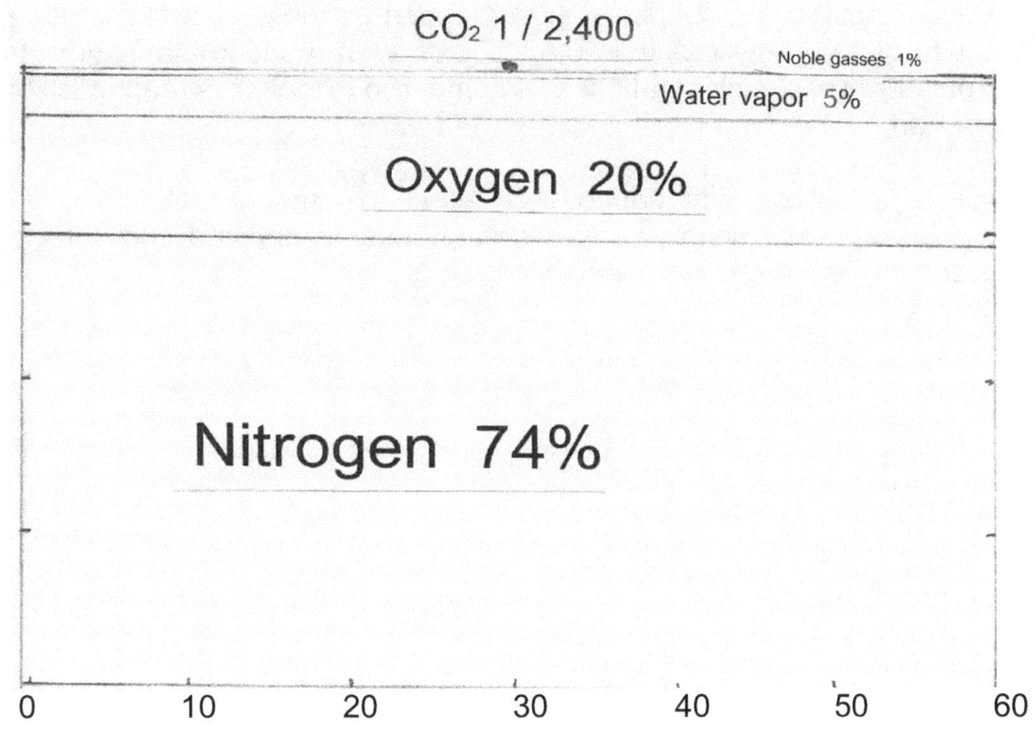

Fig 56. A graphic capture of the puny amount of CO_2 in the atmosphere. This amount equals 0.0417%, present in our air in early 2020. The ordinates, 60x40 represent the 2,400 rectangles of which one is CO_2.

In Part TWO , we went back two billion years. We discovered that, before the vegetal world began, our paleoatmosphere was mainly CO_2 and N_2 on a dry basis. That CO_2 amounted to as high as 75%, the remainder being 23% N_2, 1% SO_2 and 1% rare gasses. At this concentration of CO_2 , its greenhouse effect should have been a runaway warming to as much as $1000^{\circ}C$, scorching all possible life into oblivion. Yet paleontology teaches us that stromatolites and other early CO_2 splitting vegetation were thriving. The corresponding climate may have been 2 to $4^{\circ}C$ warmer than today but not more than that.

What does that mean?   It means, unescapably, that the IPCC has overestimated the greenhouse gas effect of CO_2 in air by a factor of 100 to 1000. Let us be conservative and settle for 200.
So, when the IPCC declares ex-cathedra that the 140 ppm rise, so far, of CO_2 has produced already $1^{\circ}C$ increase, the reality is $0.005^{\circ}C$.
When the IPCC claims we could see up to $4^{\circ}C$ rise by the end of the century, from the continuing rise of CO_2 , the reality is $0.02^{\circ}C$, not even measurable in our atmosphere. These are obviously negligible quantities as compared to the sun's magnetic contributions proposed in Part Two.

In a nutshell, CH_4 has zero effect and CO_2 's effect is unmeasurable.
We will now develop the history of that greenhouse theory and show the contortions pseudo-scientists went through to shore up their theory.

* * * * *

<u>**2. The Architects of Greenhouse: Arrhenius, Keeling and Sagan.**</u>

There were a few early proponents before Arrhenius such as Joseph Fourier, John Tyndall and Claude Pouillet but the CO_2 rise in the air had barely started. Thus it was pure speculation to assign any warming to coal burning gasses at that time, 1850-1890. More recent pioneers include Callender, Taylor and others.

Sources: Wikipedia, 1972 Encycl. Brit. for both Arrhenius and Keeling.

<u>**Svante August Arrhenius. Vik 1859 - Stokholm 1927**</u>   **Fig. 57**

A dominant scientific personality of his time, Arrhenius owed his fame to the theory of electrolysis and his discovery of the ionization of dissolved acids and bases which earned him the 1903 Nobel Prize for chemistry. He became a towering Nobel organization figure until his death in 1927, favoring friends such as Oswald, van 't Hoff and Richards and denying the prize to "enemies" such as Einstein, Angstrom, Mendeleev, Nernst and Ehrlich.

Among his many pioneering efforts in chemistry and physics, Arrhenius picked up on the Swiss and German reconstruction of ice ages in the Alps. Quantifying the cooling involved in ice ages was attempted 100 years before the actual proxy determination on Greenland and Antarctic ice cores. And it came pretty close: -7 to 10°C.
At about the same time, 1896, Arrhenius had proposed the hypothesis that "carbonic acid" from vastly increasing coal burning was raising the temperature of our planet.
Nobody had successfully analyzed CO_2 in the air so Arrhenius' calculations were based on assumed differences.

Arrhenius was first and last to assign the entire ice age drop in temperature, up to 10°C, to a decrease in CO_2 content in the air. He claimed that one half drop in CO_2 would drop the world's temperatures by 5°C. Nobody today goes that far! Just like with Carl Sagan 64 years later, Arrhenius' scientific judgment fell victim to an extreme belief that carbonic acid was in total control of our climate. The life profile of Svante Arrhenius in the 1972 Encyclopedia Britannica does not mention at all the climate issue discussed above. One more sign of the greenhouse story being recent on the world's radar.

* * * * *

Keeling receives the Medal of
Science in 2001

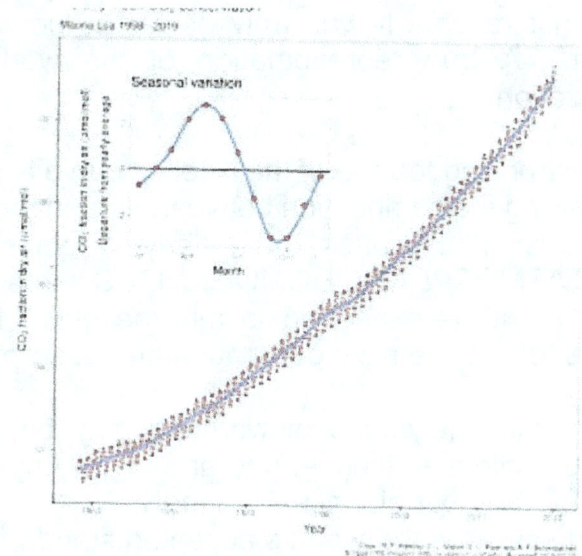

Atmospheric CO_2 concentrations measured at
Mauna Loa Observatory: The *Keeling Curve*.

Fig. 58

In the "Greenhouse effect of CO_2" camp, Dr. Keeling was the only one scientist who MEASURED and limited himself to the rise of CO_2 fact. He strongly believed that the rise of CO_2 did affect climate but soft pedaled that message unlike Arrhenius and Sagan.
Born in Scranton, PA, Charles earned a BS in Chemistry from the U. of Illinois in 1948. He was barely 20. Then, in 1953, he achieved a PhD in Organic Chemistry from North Western University under Malcolm Dole, a polymer chemist. This was a straight passport to the oil industry. However Charles Keeling had trouble seeing the future this way. Early in his PhD work, he had strongly reoriented himself towards Geology and acquired most of the credits available. So much so that when applying for his first job, West of the Rockies, it was all in Geology departments. Landing at Cal Tech under Harrison Brown, he joined in 1956 the Scripps Institution of Oceanography where he stayed to the end. In addition, he was named Professor of Oceanography at that Institute in 1968.

Keeling was a true pioneer: at Cal Tech, he soon developed (1956 -1957) the first accurate instrument to measure carbon dioxide concentrations in air samples. To test his device, he camped at Big Sur on the California Coast and guessed early that CO_2 had risen since the 19th Century. Thanks to the Scripps Director Roger Revelles connections, Keeling received in 1958 the key funding to build the famous CO_2 sampling and analyzing laboratory at Mauna Loa on the Big Island of Hawaii. By 1960, Keeling had firmly established the seasonal variation of CO_2 balancing vegetal absorption with human production and the small progress of carbon dioxide in our world atmosphere, 0.5 to 1.5ppm per year.

At 13,680 ft altitude, in the middle of the Pacific Ocean and despite occasional short-lived CO_2 emissions from nearby volcanic eruptions, the Keeling lab's location was a master stroke of inspiration. It was truly as far as possible away from CO_2 generation and absorption. A true representation of the world's tropospheric air after complete homogenization.

The most surprising feature of the Keeling curve is its consistent, yearly sinusoidal shape clearly showing the competition between human production and vegetal absorption. Why is it so surprising? Because, the underlying requirement is the SPEED AT WHICH THE WORLD'S ATMOSPHERE HOMOGENIZES across its entire surface. Only a few weeks, at most a month, are needed to mix the entire troposphere so completely that ppm differences reflect chemical generation and absorption of CO_2 with virtually no delay.

No matter what the cause of warming may be, Charles David Keeling was a giant contributor to climate science. No arguing, no hypothesizing , just exquisitely accurate analysis of CO_2 in our global air! Period!
The Keeling curve is the world's accepted standard for CO_2 content in air. Nothing else comes close or can even compare!

* * * * *

<u>4.   **Carl Edward Sagan. Brooklyn, N.Y. 1934, Seattle 1996.**</u>

Fig.59   From Smithsonian, March 2014, entitled: "Star Power". Art Work. Judy Hewgill. [29]

The epitome of a successful science popularizer, Sagan was primarily a dreamer: Finding other intelligent life in the Universe was the goal of his life and career. When it came to true science, however, Carl did not shine as well as with his active imagination, almost as a magician. He had a rare gift for mesmerizing his audience.

Sagan attended the University of Chicago where he earned a Bachelor (1955) and a Master degree (1956) in physics , and a PhD (June 1960), in astronomy and astrophysics. Carl Sagan focused his early work on the atmosphere of the planets Jupiter, Mars and most specifically Venus. The publication that followed his PhD thesis and summarizes it 9 months later is:

The Planet VENUS: Recent observations shed light on the atmosphere, surface and possible biology of the nearest planet. Carl Sagan.[30]
SCIENCE, 24 March 1961, Vol. 133, N° 3456, pp 849-856.

Two more papers followed on the subject when he was at MIT and while he already focused on more distant astronomy:
The Infrared Limb Darkening of Venus. By James B. Pollack and Carl Sagan.[31]
Journal of Geophysical Research, Sept 15, 1965, N° 18, pp 4403-4426.
Pollack: Harvard College Observatory, Cambridge , Mass.
Sagan: Harvard University and Smithsonian Astrophysical Observatory, Cambridge,Mass.
and Alfred P. Sloan Foundation Research Fellow.
Anisotropic Nonconservative Scattering and the Clouds of Venus. By Carl Sagan and James B. Pollack [32]
Journal of Geophysical Research, Jan 15, 1967, Vol. 72 N° 2, pp 469-477.

This volume of publications makes it obvious that Sagan researched the atmosphere of Venus pretty extensively. Less obvious from the titles is the deep underlying thesis: it is the Runaway greenhouse gas effect of the CO_2 atmosphere of Venus that makes it so hot." (700° – 900° F).

The thesis was based on very shaky evidence to the point of being nearly rejected. It was a good friend that saved the day for Sagan.

The following picture of Venus clearly shows how thick clouding permanently hides the hard surface of the planet. So, no possibility of infrared radiation back to space through these heavy clouds.

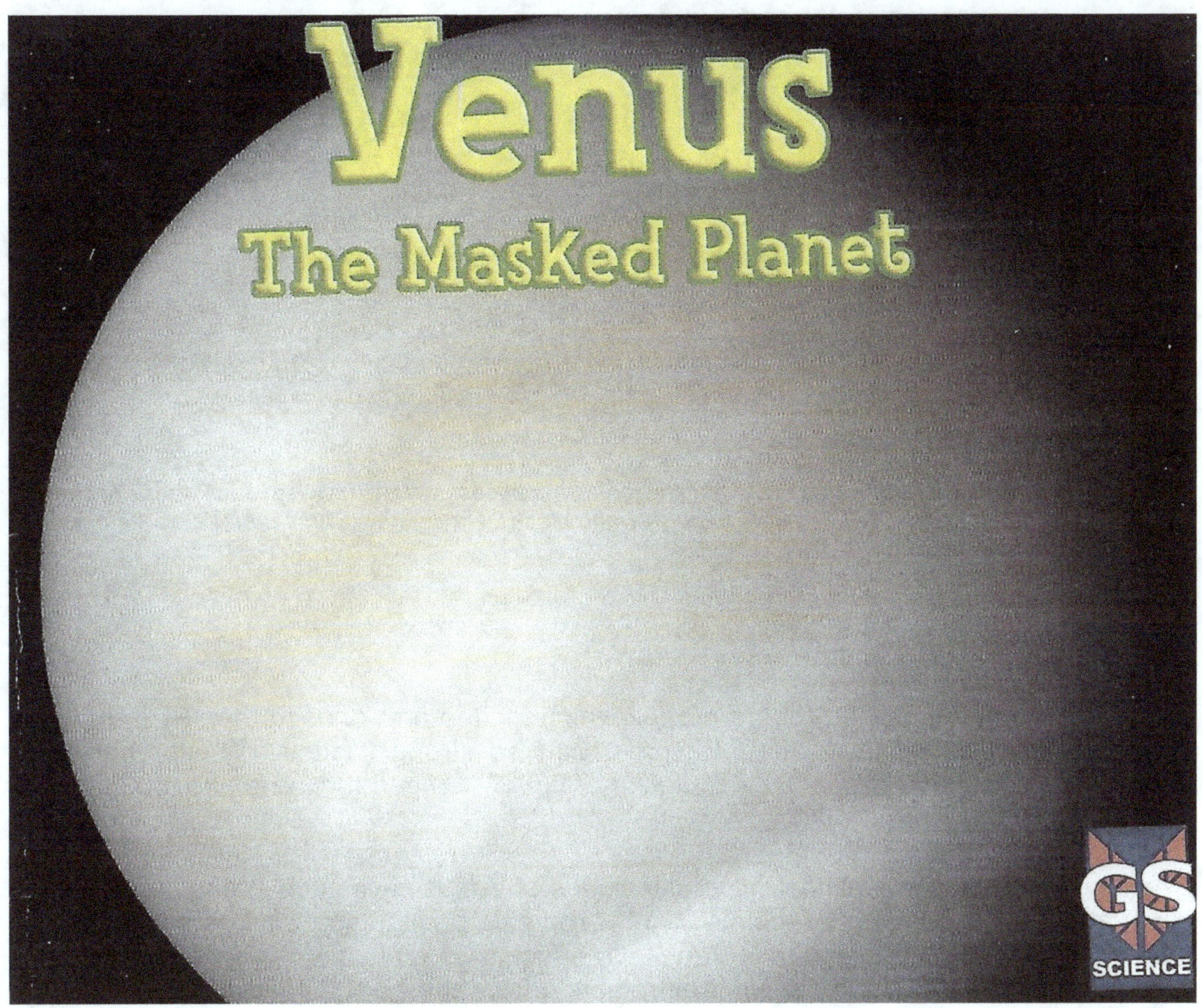

Fig 60. A recent photograph of Venus clearly showing its completely clouded surface. Venus is closer to the sun than Earth. Thus this fully sunlit picture is taken when Venus sits very far away from earth, almost exactly with the sun behind the earth. The light blondish color of the clouds indicate that most of the sun's light is radiated back to space by the albedo effect. No greenhouse gas effect whatsoever! Lincoln James, the author of the booklet, never mentions the nature of the clouds: volcanic ashes kept aloft by heavy winds and dense atmosphere.[33]

So, what is masking the hard surface of Venus?
We know that these clouds contain virtually no water. We have silica in the spectral analysis and we learned more recently that Venus sports some 1500 volcanoes, many of them still active. So the clouds, with their silica signature are volcanic ash. Remember Mt Pinatubo in the Philippines with its massive ash cloud? Well, on Venus, the very dense CO_2 atmosphere keeps similar ash clouds almost permanently aloft, aided by heavy winds as the pictures of the planet suggests.

Fig 61. A very popular radar picture of one of the major volcanoes on Venus, called Maat Mons. The colors are not real. The reality @865°F is very dark gray and black. Almost no sunlight penetrates the clouds to reach the hard surface.[34]

At least one of the Soviet Venera probes which landed on Venus in the 1970s touched down on a solidified lava lake. Pictures required strong flashes. The probes were designed to survive some 20 minutes the 865°F temperature and the 100psi CO_2 atmosphere.

As a consequence, the almost white ash clouds of Venus have such strong albedo effect that about 80% of he sun's incoming light is radiated back to space. The earth absorbs far more of the sun's energy than Venus. In effect, thus, Venus is hot because the thick, dense volcanic ash clouds prevent all radiation from the hard surface and insulate the volcanic surface with an impenetrable blanket.

The only heat escaping to the upper clouds for radiation to space is by <u>convection through the atmosphere</u>. That is why Venus is so hot: Heavily blanketed to keep warm!

Back to Carl Sagan! The modern picture of Venus totally cancels any greenhouse gas effect of CO_2. Yet the image is so deeply entrenched in peoples minds that all references to Venus still insist on the "runaway greenhouse effect of CO_2" on Venus.

In "***Climate Change: Natural or Manmade?***" Joe Fone[35] refers to a scandalous book titled "***Worlds in Collision***" by Immanuel Velikovsky. This book explained the high temperatures of Venus by proposing that the planet was billion of years younger than Earth and Mars. It created an uproar from most astronomers including Sagan. It is even probable that it drove Sagan to explain the heat on Venus with a competing mechanism, the greenhouse effect.

Around 1968 I remember reading an article in a popular science magazine in which Carl Sagan explained his greenhouse theory. I was taken aback by the shallowness of his arguments and his tentative application of it on earth's atmosphere with 0.03% CO_2 at the time.

In 1972, Carl Sagan and George Mullen published a fourth paper[36] extending the greenhouse gas effect of CO_2 to Mars and Earth.
The title of that paper is: *"**Earth and Mars: Evolution of Atmospheres and Surface Temperatures.**"* This is among the strangest scientific work I have ever read for the following reasons:

   a) Three and two billion years ago, the sun would have been 40 to 60% weaker than today!   Completely at odds with paleontological evidence of normal temperatures. The authors cite a flurry of scientific references which look today like a fashion of the 60s. The claim was based on young stars of our galaxy being all weaker than mature stars. How could this be reliably determined? Mind boggling! The net result, from brief energy computations, is that our planet would have looked like a snowball. The authors indeed struggle with any chance of life existing at all unless a greenhouse Deus ex Machina could be called in. But aside from ppms of NH_3 and CH_4 they could imagine being present, they could not find any big greenhouse actor.

   b) At that same remote time, they claim - hard to imagine - that "CO_2 concentration was about the same as today", roughly 0.035%. This, a complete repudiation or ignorance of the now solid fact that all of our oxygen today comes from the splitting of ancestral CO_2 by vegetal life as developed in Part Two.

   c) "Contrary to geological and paleontological evidence"... This statement appears early in the abstract and refers to the first two points a) and b). By that very statement the authors took a serious risk of irrelevance in all what they advanced in the paper.

Of course, very few people read those scientific articles, and Sagan had a nag of making them particularly hard to read using for example aeons instead of billions of years. This stratagem allowed Carl Sagan to baffle his unscientific audience with the new greenhouse theory presented as a fact.

Venus is not hot from "runaway greenhouse effect". There must be another origin to that heat which manifests itself by an abundance of volcanic activity on the surface.
On the other hand, Venus shows almost no magnetic field like the earth does. A possible source of the high volcanic activity may be a high quantity of nuclear fissile material deposits deep into the core of Venus: Uranium, Thorium, Radium, etc. which act as several major heat sources over the billions of years of the life of the planet. If true, this fissile material replaces much of the iron, nickel and iridium making up the center of our planet and generating its magnetic field.

Thus Venus could be hot because of higher fissile elements content than the earth. That energy content would not be inexhaustible but could last billions of years before playing itself out slowly. This conveniently replaces Velikovsky's young planet approach and rejoins Taylor's interpretation of the cause of high heat on Venus.

Finally, in late 1975 or 1976, Sagan, a very persuasive orator, was invited to present his climate theory to the US Senate. This is how, young freshman Senator Albert Gore was forever converted to the CO_2 greenhouse theory as the cause of global warming.   This theory was unheard of in the mid-seventies.

By 1980, the fiction of the greenhouse gas effect of CO_2 had grown deep roots in the U.S. and Europe just in time to create a magic alliance with the Green movement of Sweden and Germany.

* * * * *

<u>**5. In 1980, the Green Movement Stumbles Onto Sagan's Fiction: Perfect Match!**</u>

The original Green movement accused INDUSTRY for being the source of all societal problems in developed countries. At the limit, they advocated a "return to cave man" which, of course, weakened their political appeal. In search of facts to support their anti-industry position, the Green Movement stumbled around 1980 on 3 new concepts:

1/ New analysis of CO_2 in our air showed increase: The Mauna Loa Lab and the Keeling curve.

2/ Increasing temperatures of our globe: Global Warming barely perceptible in 1980.

3/ The greenhouse gas effect of CO_2. The fiction that connected 1/ and 2/. Carl Sagan's cat was out of the bag for good!

The first point, the rise of CO_2 was undeniably caused by our fossil fuel burning industry. The very vocal Green Representatives made sure that it proved that our woes started with the "Industrial Revolution" around 1850.

The second point, the warming was more recent after an apparent cooling in the late sixties and seventies. However, during the eighties, it became fairly convincing that the warming of our globe was accelerating.

The third point, strongly supported by a very popular scientist, Carl Sagan, was essentially swallowed as fact by the Green movement, the media and the politicians on the left.

* * * * *

6. 1980-88. The Ascent of Man's Guilt as a Religious Belief.

It was the underlying current of suspicion against industry that the Green movement had planted which took hold in Society. It was, of course, less extreme than the Greens wanted but it solidified around the greenhouse gas effect of CO_2. The extreme political position of the Green movement had subsided. They had initiated the issue and it was now in the general public through the media. Those media loved the subject. It gave them the excuse to go around the world to prove warming, particularly around the poles. Then, they always concluded the report saying "and, of course, it is CO_2".

The exponential look of the Keeling CO_2 curve, encouraged the birth of catastrophism: If we don't do anything, climate is going to "kill the planet". "Got to stop the CO_2 rise!"
This growing popularity of urgency against "Global Warming" translated into "finding a global way to solve the problem". Many possible platforms were looked at during the mid eighties by young self appointed "Climatologists". The first attempts to scientifically support the greenhouse theory began also in the eighties with computer modeling of weather, etc.. The difficulty of grasping all factors affecting climate became quickly apparent. It advocated strongly for a worldwide effort and concurrent budget for research.

The first schoolbooks appeared also with the subject of "Global Warming" never questioning the validity of the greenhouse gas theory or suggesting any alternative cause.

Towards the end of the decade, 1986-87, it was finally obvious that the United Nations were sufficiently recognized by the world to assume the "climate responsibility" in a new department.

* * * * *

7. 1988-98. The Master Plan to Prove Man Guilty. The IPCC.
The Intergovernment Panel for Climate Change.

Among movers, Al Gore and Jim Hansen of the U.S., Maurice Strong of Canada and Phil Jones of East Anglia, U.K. were highly instrumental in setting up a whole new department at the United Nations exclusively in charge of climate. The well documented purpose of the Panel was to analyze and study climate data and climate factors in order to curb the rising temperature of our globe.

The fundamental platform of the new IPCC at the U.N. was preset: The Greenhouse Gas Effect of CO_2 IS THE CAUSE of Climate Change.

The entire purpose of the IPCC was to scientifically shore up that predetermination as unquestionable fact. Research budgets were being set up to finance huge computer models, worldwide temperature recordings in new settings, satellite programs, etc...
Early optimism was almost boundless and international support grew from European, Canadian, Japanese, Australian and U.S. entities. Universities such as East Anglia in the U.K., M.I.T. and Penn State in the U.S. were early recipients of generous U.N. funds which extended later to many others. Climate Change careers sprang overnight creating a new respected status of "Climatologists".
There was still some genuine honesty in the first official IPCC report. The perfect example is the following diagram:

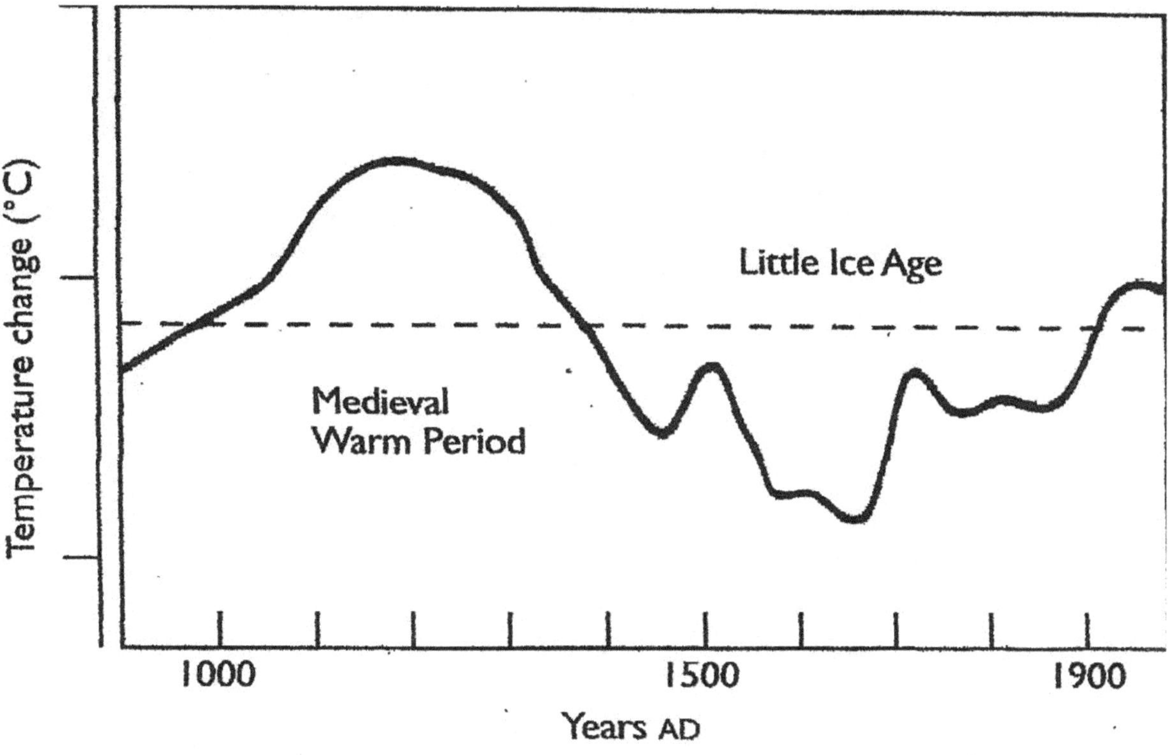

Fig 62. An amazingly accurate temperature-time diagram from the first official IPCC report in 1990. This cancels any possible effect of CO_2 on our climate and was soon disowned and abandoned by the top IPCC staff.

Soon after publication, this diagram was recognized by the most astute directors of the IPCC as a major "faux pas", a blunder of enormous consequence. Indeed, they pointed out secretly, if one admits a previous warming one thousand years ago, it cancels the rise of CO_2 since the Industrial Revolution as the cause of warming today. In a secret meeting sometime around 1991, it was firmly decided never to show any historic climate diagram in the future. And it stuck!

Among emails revealed by the 2008 secret email debacle, perhaps the most significant was:

"We have got to get rid of Medieval Warmth and Little Ice Age…"

This short secret sentence by an eminent "Grand Father" of the CO_2 fiction at the IPCC demonstrates his profound dishonesty of purpose: <u>Erasing historic climate science to shore up CO_2 fiction.</u> In fact he demonstrated by that short sentence that he knew that CO_2 was a fake!

* * * * *

8. 1998-99: Denying Historic Warm Middle Ages: The Hockey Stick.

"We have got to get rid of medieval warmth and little Ice Ages"!   In that same sentence, an order was given to new climatologists: Someone better comes up with a "scientific proof" that temperatures did not rise in the Middle Ages and dip from 1300 to 1700. That order took from 1991 to 1998 to be executed. At Massachusetts Institute of Technology, a young team led by Michael E. Mann went to work with the sole purpose of erasing medieval warmth and mid-millennium cool.

The first paper, published in NATURE in 1998, only went back to about 1400 AD to erase the Little Ice Ages.

Mann M.E., Bradley R.S., Hughes M.K.:[37] *"Global-scale temperature patterns and climate forcing over the past six centuries".* Nature, 1998, 392, pp 779-787.

Based on cherry-picked data from ancient bristle cone pine rings, Mann et al. <u>claimed the absence of cooling between 1400 and 1700.</u> The title itself of the paper was neutral but the content was clear.

Then a second paper extended the deception to one thousand years back:

Mann M.E., Bradley R.S., Hughes M.K.:[38] *"Northern Hemisphere temperatures during the last millennium: inferences, uncertainties and limitations".* Geophysical Research Letters, 1999, 26, p 759-62.

That second paper executed the 1991 command, to complete satisfaction for the IPCC. It was a "tour de force" to cherry pick proxy data to the point of erasing completely the strong historic medieval warming of the order of 2 to 2½°C. That temperature evidence comes from both Greenland and Antarctic ice cores, comparing O_{16} to O_{18} isotopes. It was, of course, available to the Mann's team but to a zero effect.

The diagram of that second paper was nick-named: "The Hockey stick".

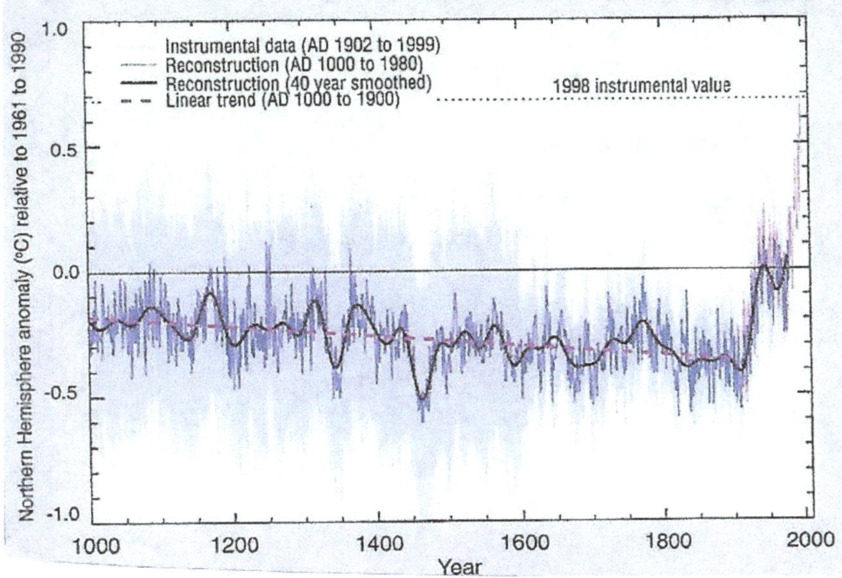

Fig 63   Michael Mann's "Hockey Stick" temperature reconstruction as it appeared in the IPCC Third assessment report of 2001 showing fluctuations in relation to the 1961-1999 mean temperature. Data from instrumental temperature records starting in 1902 is overlaid in red over estimates based on climate proxies (tree rings, corals, ice cores, etc..) from the last 1000 years. The grey area represents estimated error levels. From the back cover of the 2010 book by A.W. Montford.[39]

That 2010 book by A.W. Montford is titled: "The Hockey Stick Illusion".[39] The Scottish team work is largely based on intensive inquiries by Mc Intyre and Mc Kitrick into the convoluted network of Michael Mann's "Defense Zone". The following table from the same book shows an amazing amount of publications supporting the Hockey Stick.
All this pseudo-science displays an impressive array of proxies intended to negate the historic truth of warming and cooling cycles of the last 1000 years

TABLE 10.1: Commonality of proxies in temperature reconstructions

Proxy series	MBH98 & MBH99	Rutherford	Jones 98	Crowley 00	Briffa 00	Esper 02	Mann, Jones 03	Moberg	Osborn, Briffa	D'Arrigo
Polar Urals	x	x	x	x	x	x	x	x	x	x
Tornetrask	x	x	x	x	x	x	x	x	x	x
Jacoby Mongolia	x	x			x	x	x	x	x	x
Jacoby treeline	x	x	x	x	x					x
Bristlecones	x	x		x		x	x	x	x	
Dunde/Yang	x	x		x		x	x	x	x	
Greenland $\delta^{18}O$	x	x	x	x			x		x	x
Jasper			x	x	x	x			x	x
Taimyr					x	x	x	x	x	x
European docs	x	x	x						x	
CETR	x	x	x	x						

Adapted from the Wegman report.[15]

Fig 64. The first publication entry is abbreviated from Mann, Bradley and Hughes, 1998-99. Jones is for Phil Jones from East Anglia. So are Briffa and Osborn.

The Official 2001 adoption by the IPCC of Mann' Hockey Stick papers as new standard for 1000 years of paleo-climate generated profound skepticism mostly among older scientists who knew better than "Greenhouse" fiction. Mann's reaction is reminiscent of the circling of the wagons by South African Boers to resist Zulu attacks. The following figure illustrates that reference, still from Montford's book.[39]

FIGURE 9.2: Mann at the the centre of a paleoclimate web

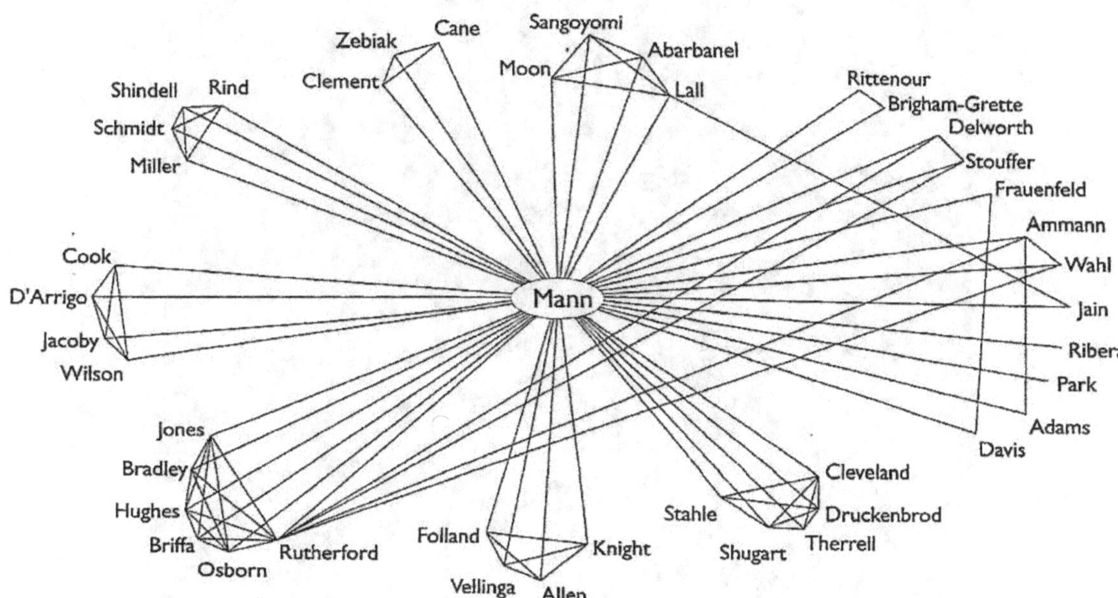

Lines indicate links of co-authorship. While each clique is largely self-contained, Mann has worked with all of the authors shown. Adapted from The Wegman Report

Fig 65   Michael E. Mann had a genial gift for surrounding himself with a powerful array of circling wagons to defend his pseudo-science.

This defense strategy worked with the strong support of the IPCC top brass to which some of them ascended in the first decade of the 21st century. The older scientific "skeptics" were out maneuvered.

By 2013, Michael Mann had the guts to publish his own book [40] brazenly challenging the opponents to his Hockey Stick deception.

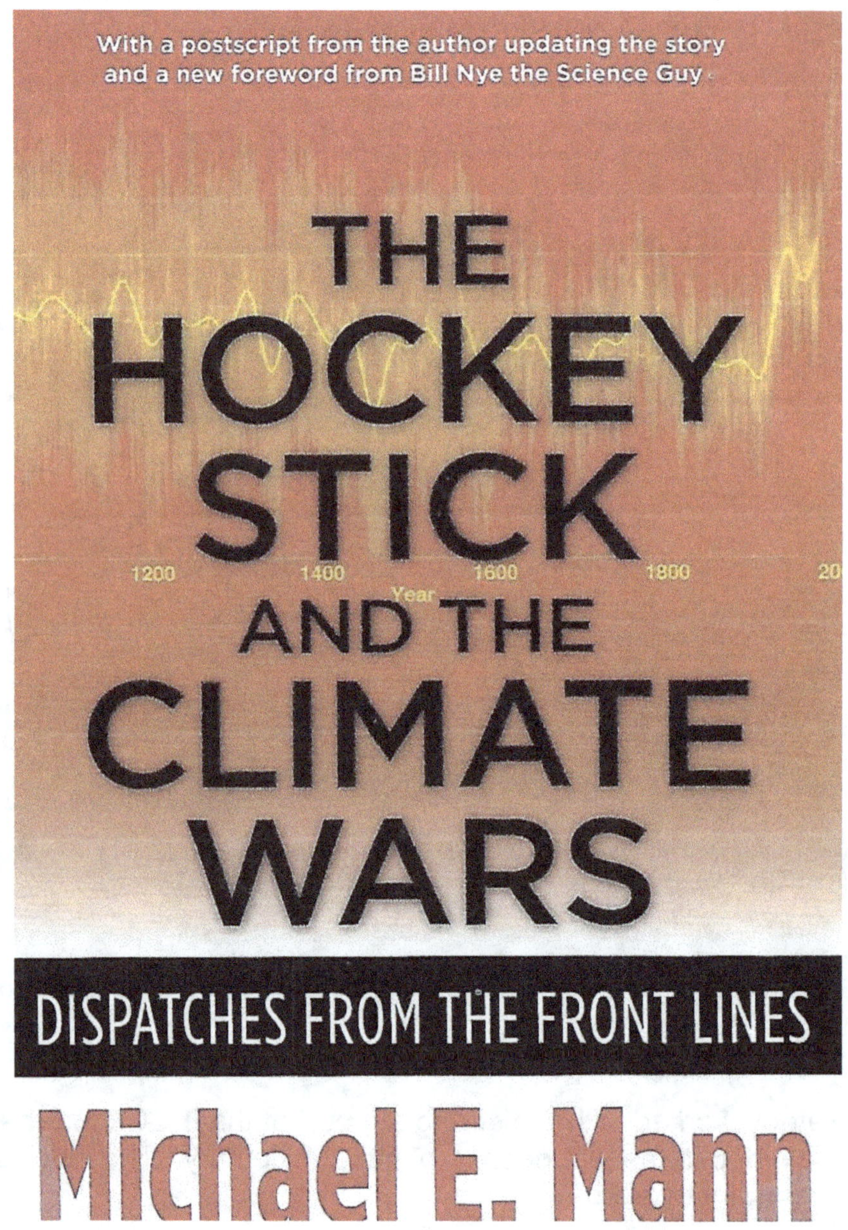

Fig 66. Glorifying his winning war against disorganized skeptics, Mann, the leader of the CO_2 Hoax, solidifies the IPCC climate campaign against the fossil burning industry.

* * * * *

<u>**9. 2000-2020. Disorganized Skeptics Wage a Late and Loosing War.**</u>

There is one historic precedent to today's intense climate war: **The Crusades.**

Around 1095 AD, Pope Urban II traveled all the way to Northern France. With eloquent and compelling speeches, he inflamed his Frankish believers to paint big red crosses on their front and back. They were to leave wife and property at home, cross Europe and "SAVE" Jerusalem, the Holy City, from the pagan muslims by force of arms.
This was profoundly unchristian: Christ's message upside down! Ironically, it was, in fact, the muslim concept of converting people by conquest and force applied by a Christian Pope.
There must have been good Christians and, particularly, well educated monks who were skeptic of this crusade call. However, even think of fighting your own top boss and you were certain of excommunication from the church. No wonder we have no chronicle evidence of crusade skeptics!

This example illustrates the religious aspect coloring the climate war: "Earth in the Balance" by Vice President Al Gore [41] in 1992 followed by the same author of "Inconvenient Truth" [42] in 2006-2008. Save the Planet reminds us of Saving Jerusalem.

<u>**The CO_2 Hoax and Skeptics: A Tentative Time Line.**</u>

1990. The first, clear evidence of intention to prove man guilty of climate change came in 1990 from the first official IPCC report.   Then, on an accelerated pace and worldwide, the media tuned in and amplified this United Nation message, automatically international. Al Gore published his 1992 book "The Earth in Balance". [41]. The Climate Research Unit (CRU) at East Anglia University in the U.K., with Phil Jones et al. in charge, contributed articles and ad hoc research, pioneering the future climate programs at Penn State, MIT, etc. Jim Hansen, friend of Al Gore, took care of keeping Congress, particularly the Senate, updated on the progress of the new wave focusing on CO_2 as major culprit for warming. It was increasingly political.

1992. Rio de Janeiro: The first climate conference.

1995. A critical "scientific" vote was conducted by the IPCC among 25 scientists supposed to know something about climate: The question was: Does CO_2 influence climate? Yes or No?
The "Yes" won by one vote, 13 to 12 and the IPCC trumpeted their victory. This narrow victory for CO_2 shows, however, that there was considerable doubt, at the time, about the greenhouse gas theory of Carl Sagan. With 20/20 hindsight, it would have been a key milestone for the 12 "Negatives" to issue a statement in the press as early skeptics of this new wave. Instead, it was total silence.

1997. Kyoto, the second, most successful Climate Conference but without even denting the rise of CO_2 .

1998-1999. [37,38] Mann et al. publish the Hockey Stick papers as previously detailed.

2003. The first skeptic reactions to the Hockey Stick of Mann et al. have to wait to that that year, most notably:

Mc Intyre S. and Mc Kitrick R.[43] *"Corrections to the Mann et al. (1998) proxy data base and Northern Hemisphere average temperature series".* Energy & Environment 2003. 14, 751-771.

--- We have to wait another 4 years for the first skeptic books. ---

#1. 2007. *"The Politically Incorrect Guide to Global Warming..."* Christopher C. Horner.[44]

#2. 2007. *"Unstoppable Global Warming Every 1500 Years."* By S. Fred Singer and Dennis T. Avery. [27] Unfortunately, the 1500 year cycle has not been confirmed by any other skeptic so far. It is based on cycles during the major glaciations, not on Holocene cycles. The latter are 1000 year cycles.

--- Another 3 years and, finally, the Skeptics flood gate opens. ---

#3. 2010. *"The Hockey Stick Illusion. Climategate and the Corruption of Science"* by A.W. Montford.[39] . Largely based on the 2003 Mc Intyre & Mc Kitrick paper. A milestone to which I refer frequently. Of particular interest is the last chapter, p. 400: The CRU Hack: The East Anglia email trove penetrated by skeptic hackers. It reveals the inner machinations of Jones, Mann et al.

#4. 2010. *"Climate Gate: A Veteran Meteorologist Exposes the Global Warming Scam."* by Brian Sussman[45] (San Francisco TV)

#5. 2010. *"Climate of Extremes: Global Warming Science They Don't want you to Know"* by Patrick J. Michaels and Robert C. Balling, Jr.[46] Soft skeptics confirming the 20th century warming.

#6. 2010. *"Climate: The Great Delusion. A Study of the Climatic, Economic and Political Unrealities".* By Christian Gerondeau.[47] Preface by Valery Giscard d'Estaing. A French skeptic book translated in English.

-- Despite missing some possibly important skeptics books, we can tell from the preceding and following list that we are in the middle of a skeptic peak --

#7. 2011. *"Cold Sun: A Dangerous Hibernation of the Sun has Begun"* John L. Casey.[23]

This prediction of the "Second Dalton Cooling" is one of the basic references for point 11 of PART TWO. There is now mounting evidence that Casey's prophecy was a decade early. Climate hysteresis is at work here as explained on that point 11.

#8. 2011. *"Climate Coup: Global Warming's Invasion of our Government and our Lives"* Edited by Patrick J. Michaels.[48] Authors include Roger Pilon, Evan Turgeon, Ross Mc Kitrick, Ivan Eland, Sallid James, Indur M. Goklany, Robert E. Davis and Neil Mc Cluskey. Focused on the climate turn-around by President Obama.

#9. 2011. *"Don't Sell Your Coat! Surprising Truths About Climate Change"*. By Harold Ambler.[17] Good reference for both Parts 1 and 2 of this book: Abdussamatov, Eddy, Mann, Gore, Svensmark, Livingston, etc.

#10. 2011. *The Delinquent Teenager Who was Mistaken For The World's Top Climate Expert"*. IPCC Expose by Donna Laframboise.[49] A devastating report on the profound dishonesty of the United Nations Intergovernmental Panel on Climate Change.

#11. 2012. *"The Greatest Hoax: How the Global Warming Conspiracy Threatens Your Future"*. By U.S. Senator James Inhofe.[50] The politics of the climate issue exposing the power grab of the left through carbon taxes and a growing climate budget based on faulty science.

#12. 2012. *"The Great Global Warming Blunder. How Mother Nature Fooled the World's top Climate Scientists"*. By Roy W. Spencer.[51]

#13. 2012. *"Hiding the Decline: A History of the Climate Gate Affair"*. By A.W. Montford.[52] This second book after the first 2010 issue updates us to the latest IPCC "tricks" particularly in hiding global temperature leveling after 1998 and decline through 2011.

#14. 2013. *"The Neglected Sun" Why the Sun Precludes Climate Catastrophe"* By Fritz Vahrenholt and Sebastian Lüning. [25] The original German version: ***"Die Kalte Sonne"*** dates from early 2012. This was a major source for this book including the reference to Jack Eddy, the 1000 year cycle, the temperature proxies, the 21st Century predictions, etc. It was also the first book clearly pointing to the sunspot cycles as cause of warming and cooling.

#15. 2013. *"Climate Change: Natural or Manmade"* By Joe Fone.[35] Contains a very instructive chapter on Carl Sagan's greenhouse fiction.

#16. 2014. *"Warming? Yes! Manmade? No!"* By Leon A. Luyckx.[1] Published on Amazon. On page 43, the direct linkage between polar lights and global warming from the poles is proposed for the first time: direct transfer of magnetic energy from sunspots morphed to heat on our poles.

#17. 2014. *"The Deliberate Corruption of Climate Science"* by Tim Ball, PhD.[53] Another well researched and convincing condemnation of the criminal IPCC manipulation of data to prove man guilty.

#18. 2014.*"Dark Winter: How the Sun is Causing a 30year Cold Spell"*. John L. Casey.[22]

This second book on the same subject, the second Dalton Minimum, refines the predictions of the first 2010 publication but does not postpone the onset of the cooling.

#19. 2015. *"A Disgrace to the Profession.* The World's Scientists - in their own words-on Michael E. Mann, his Hockey Stick and their Damage to Science". Volume 1 Compiled and Edited by Mark Steyn.[54]The book originates from Quebec, Canada.

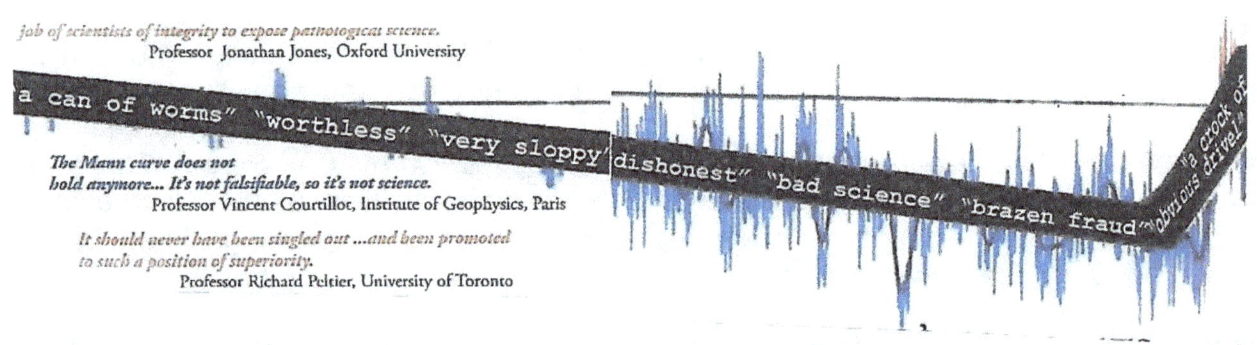

Fig 67. From the book's Front & Back Jacket: The Hockey Stick

Adding their contribution: Jonathan Jones, Prof. Oxford Univ.; Vincent Courtillot, Prof. Institute of Geophysics, Paris; Richard Peltier, Prof. U. of Toronto; Ian Plimer, Prof. U. of Melbourne and Judith Curry, Prof. Georgia Institute of Technology.

#20. 2015.*"A Climate of Crisis: America in the Age of Environmentalism"*. Patrick Allitt[55]

Most of the book is on the catastrophic views of today's environment politics. The one Chapter 9 is on global warming, summarizing the climate politics of the 1980ies and 90ies. Skeptic work, particularly by Fred Singer and Danish statistician Bjorn Lomborg is described in detail.

#21. 2015. *"A New Little Ice Age Has Started. How to survive and Prosper During the*

Next 50 Difficult Years". By Lawrence E. Pierce[56] of British Columbia. As the extreme right of the diagram below predicts, Pierce goes even further than Casey, Vahrenholt and Abdussamatov in their predictions: Not just a Dalton Cooling but a full blown little ice age.

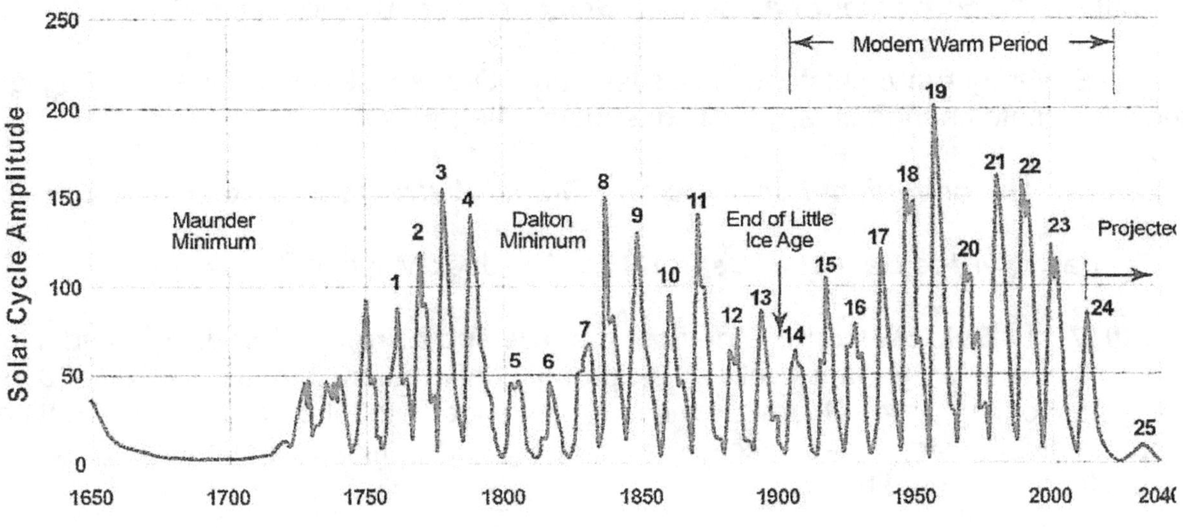

Solar Activity Cycles 1650 - 2040

Fig 68. This author calls "End of Little Ice Age" what should be named the "Industrial Minimum". Peak 25 is projected below 20 sunspots, lower than all other authors.

Pierce [56] is also the only one that I know, implicating the alignment of the planets in sunspots activity. In this, he follows Sir William Herschel, Britain's Royal Astronomer who suggested that linkage in 1801. Finally Pierce also points to volcanic activity as major climate factor, following hereby several other authors.

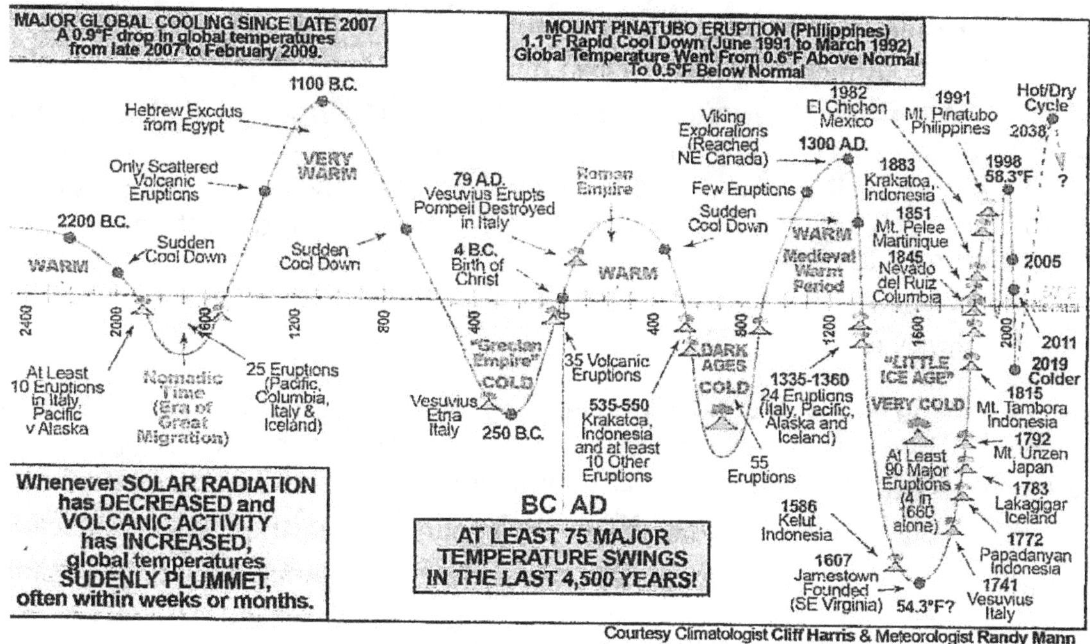

Global Temperatures 2500 B.C. - 2040 A.D.

Fig.69 A 4500 year temperature-time diagram in which Pierce sketches out many volcanic periods placed in the climate cycles. His 1100BC peak corresponds to what other authors name the Minoan maximum, universally recognized as one of the warmest Holocene periods. However, most other sources put the Roman peak temperature between the Minoan and the Medieval maximum.

#22. 2017. *"Inconvenient Facts. The Science that Al Gore doesn't Want You to Know"*

By Gregory Wrightstone.[57] Focuses on CO_2, its history, benefits and negligible effect on climate change. Many well constructed diagrams.

#23. 2017. *"Inconvenient Facts Proving Global Warming is a Hoax. The Common Sense*

Facts ***for the Basket of Deplorables"*** by Jack Madden.[58]

#24. 2017. *"A Cold Welcome. The little Ice Age and Europe's Encounter with North America"*. By Sam White.[59] A strong confirmation that the 1492-1600 mid-millenium cooling, the Sporer minimum, extended to the American Continent. Unfortunately, the story doesn't cover the deepest part of the cooling after the Plymouth pilgrims landed on our shores.

#25. 2018. *"The REAL Inconvenient Truth. It's Warming but it's not CO_2 "*.

By M.J. Sangster.[60] PhD. The case for human-caused global warming and climate change is based on deceit, lies and manipulation.

#26. 2018. *"The Politically Incorrect Guide to Climate Change"*. by Marc Morano.[61] Well referenced reexamination of the IPCC's dishonest tactics, Mann's pseudo-science, the politics and the high cost of greenhouse gas abatement.

#27. 2018. *"The Mythology of Global Warming, Climate Change Fiction vs Scientific*

Facts." By Bruce C. Bunker.[62]

#28. 2019. *"The Solar Magnetic Cause of Climate Changes and Origin of the Ice Ages"*.

By Don J. Easterbrook.[6] Abundantly illustrated evidence of the sunspots as origin of mini and maxi climate cycles. Debunks all earth-centered theories explaining climate. Easterbrook adopts the Svensmark mechanism by which the solar magnetic field prevents galactic rays from forming clouds during high solar activities, thus more warming.

The preceding list of 28 skeptic books is, by no means, exhaustive, many more are missing. As the reader may have noticed, the vast majority in the list, 21 out of 28, focus on the greenhouse gas fiction and the pseudo-science of the IPCC.

Only the following 6 point to the sun as true cause of our climate change:

7: "Cold Sun" by John L. Casey

9: "Don't Sell your Coat" by Harold Ambler

14: "The Neglected Sun" by Fritz Vahrenholt and Sebastian Lüning

16: "Warming? Yes. Manmade? No!" by Leon A. Luyckx

17: "Dark Winter" by John L. Casey

28: "The Solar Magnetic Cause of Climate Changes and Origin of the Ice Ages"
by Don J. Easterbrook

Of these six, numbers 14, 16 and 28 offer the most advanced evidence of the solar influence and # 28 dominates the entire list with overwhelming evidence of the magnetic solar influence on our climate.

Only one, #16, posits that the magnetic energy is directly ejected by sunspots and carried by matter. On arrival, it is coaxed by the magnetic field of our planet onto the poles and converted to heat, warming us from the poles.

* * * * *

The 10 Commandments of the IPCC*.

1. Declare "Ex-Cathedra" CO_2 + CH_4 Single Cause of Warming.

2. Cherry Pick Proxy Pseudo-Science to Get Rid of Medieval Warmth and
Little Ice Ages: The Hockey Stick.

3. Deep Root CO_2 Fiction Early in All Education Systems.

4. Hoodwink Media & Rich Democracies into Religious Submission.

5. Foster Media Confusing Warming Fact with CO_2 Fiction.

6. Inflict Giant Guilt on Mankind for Burning Fossil Fuels.

7. Threaten Catastrophes if we don't "Save the Planet".

8. Excommunicate Skeptics as Big Coal, Big Oil Puppets.

9. Bar Skeptics from Scientific Press through Peer Review Control.

10. Frame the 2030 "Green New Deal" as Last Chance before Irreversible Warming, Boosting from Billions to Trillions per year the International Cost of Compliance.

*The Intergovernment Panel on Climate Change.

The first 9 of these 10 commandments summarize the entire Part Three so far. The first two were covered in great detail earlier. The first and most important one enshrines religiously the profoundly dishonest statement that man is the primary cause of our climate change. The second synthesizes the length at which pseudoscientists, such as Michael E. Mann, distorted proxy data to erase the historically unquestionable evidence of Medieval Warmth and Little Ice Ages. **The Middle Age Warm Period, in fact, erases all possibilities that CO_2 causes our warming, thus exonerating mankind.**

Commandments 3 to 7 depict the aggressive, political methods used to get the world into religious submission. There are no carrots but a lot of threatening sticks. Words like "inflict" and "catastrophes" are typically the style used to force the "vulgum pecus" the common people into submission. The media love this sensationalism.

Commandments 8 and 9 crisply frame how non-believers are to be ousted from the faithful, obedient community of anthropogenic climate adherents.

The 10th COMMANDMENT.

While #1 to #9 covered 32 years since the creation of the IPCC, #10 turns to the future folly of CO_2 abatement in this decade. The IPCC has been strongly encouraged by the international success generated by its aggressive campaign of erasing opposition. Skeptics have no political support in rich democracies. No voice! Just a few barely audible objections!

The CO_2 abatement budgets, have climbed from near zero in 1988 to billions of dollars by 2000, hundreds of billions by 2010 and, probably, around one trillion per year, worldwide in 2020.   Just look at Al Gore, the multi-billionaire!
According to Bernie Sanders and Elizabeth Warren during the winter 2019-2020 campaign, just the U.S. part of the 2030 Green New Deal would cost $33 to 40 Trillions over 10 years. The same or more would be true for the E.U., plus another big slice for Japan, Canada, Australia, Mexico, etc..
In complete unbalance, Asian countries could not possibly sign up to the Green New Deal as we will detail hereunder. It would stop their economic development dead in its track.

Some smart thinkers in the media, trying to grasp the magnitude of this endeavor, have voiced their alarm. Siphoning such huge sums of money away from good social causes has to be proven effective. Or, at least, it has to show credible signs of hitting their target if applied. What better way to find out than to look in the recent past. Has **any CO_2 abatement occurred so far** thanks to the IPCC efforts and all the climate conferences since 1988? What about the return on investment of these first trillions of dollars worldwide?

The best world-accepted way is to check the CO_2 content of our atmosphere since 1988 and further back if possible. The Keeling curve is the only globally recognized, continuing record of that CO_2 content in our air.

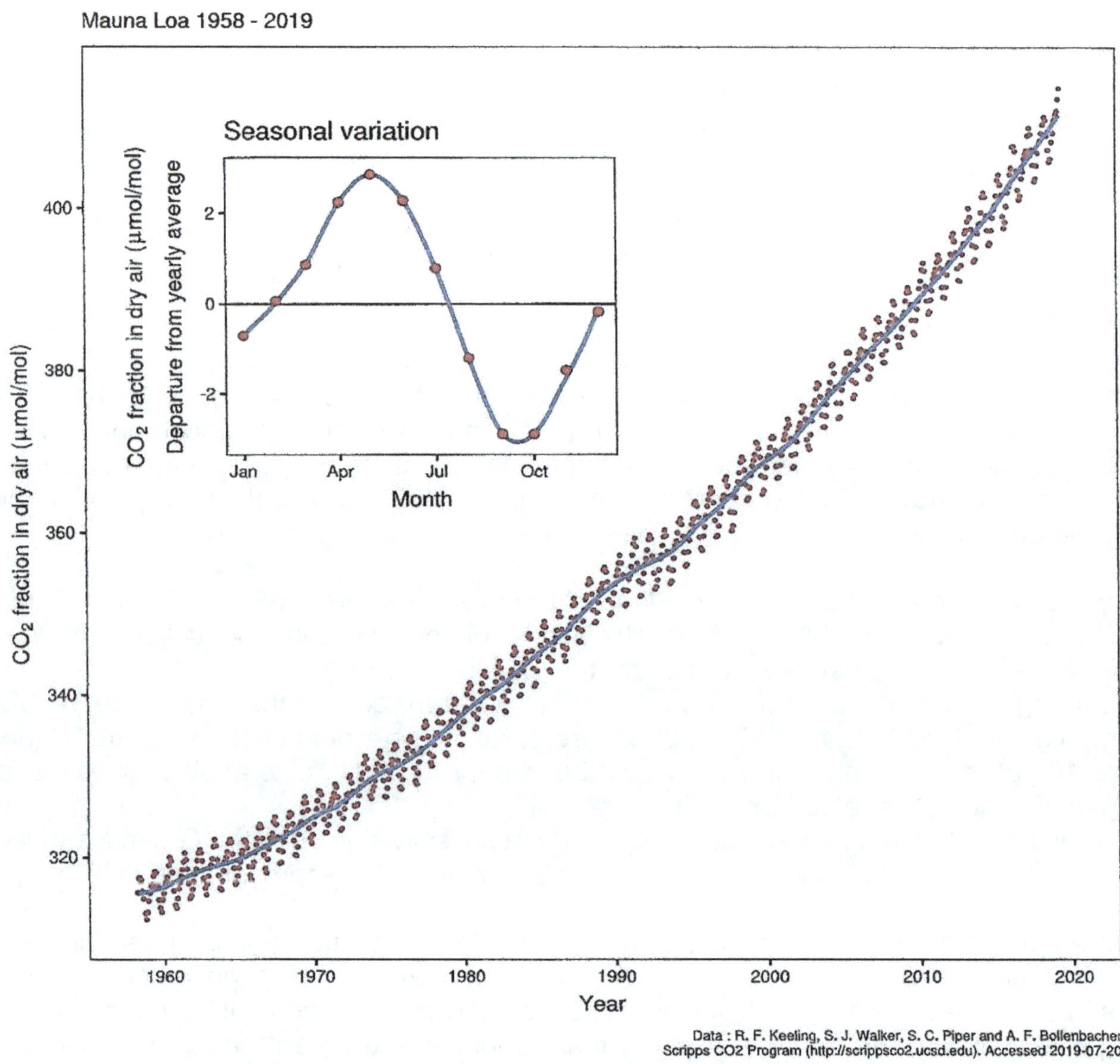

Fig.70. The Keeling curve: Every year, the CO_2 content of the world's atmosphere decreases slightly in the spring of the Northern Hemisphere because over ¾ of the vegetation sits in that hemisphere. It then increases sharply during fall and winter because the industrial generation of CO_2 vastly outpaces its absorption by the vegetation of the Southern Hemisphere. The Keeling curve is now accepted as undisputed standard for CO_2.

The first fact to point out from the Keeling curve is that **none of the CO$_2$ abatement efforts, so far have had any effect. Even the slope of the curve keeps steepening inexorably.** Not even the 2008-2009 economic crash made a blip in the curve.
While writing this book, a pronounced abatement caused by the COVID-19 pandemic may, possibly, produce the first chink in the curve. This would be good evidence of what economic disaster the 2030 Green New Deal would bring with drastic CO$_2$ abatement policies.

So, all these IPCC efforts, all these carbon taxes and transfers, the real progress made in Europe, the U.S. and Canada, have not even touched the rate of increase of CO$_2$ in the air. Hard to believe but the evidence is in the curve! You can have Kyoto, Copenhagen, Paris and Madrid! Zero Effect! It was wasted effort and money.

* * * * *

Carbon dioxide emissions did not start to rise significantly before 1850. This despite many claims that the Industrial Revolution began with the invention by James Watt of the steam engine about 100years earlier. Steel rails did not exist before 1850. **It was the steam locomotive, the iron horse, on steel rails that truly started the Industrial Revolution.** Here below, we can see how little this revolution amounted to for the first 50 years.

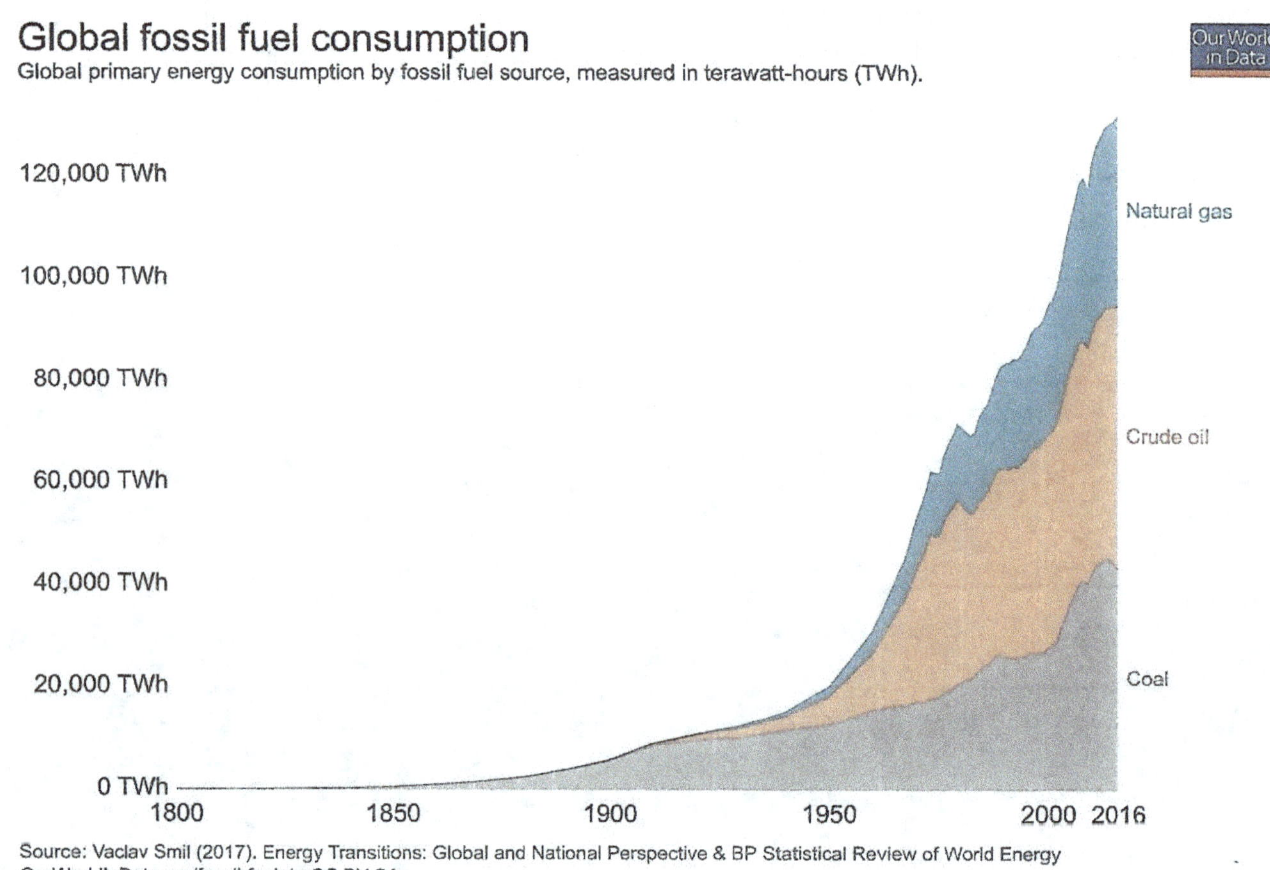

Fig. 71. In terms of Tera Watt Hours (TWh), we can clearly see how puny the impact was from the first 50 years of Industrial Revolution! The same applies to CO_2 emissions. It suggests that CO_2 was larger than 280ppm in 1850, more like 310ppm. [60]

The steam locomotive on steel rails triggered a massive surge in rail production which, by 1875, was 80% or more of steel "manufacturing". Then, by 1890, Thomas Edison started steam powered DC electricity plants for his carbon fiber light bulb.
By 1900, as the curb of fig.71 shows, coal burning now "explodes" into concrete buildings, the application of steel to construction, railroad stock and automobile manufacturing.
The plate glass industry also starts consuming coal and ship propulsion by gigantic coal fired steam engines peaks by 1912 before switching to bunker oil and Diesel in WWI. This is almost all happening in Europe, U.S., Canada, Australia, Japan, Russia and S. Africa until the 1980s.

The Asian awakening is triggered by the Great March of Mao Tse Tung in 1948. However it is only after 30 years of terrible errors by the aging and inflexible leader that China's Industrial Revolution takes hold. From 1978 thanks to Deng Xiaoping, China explodes onto the world's industrial scene. To energize this explosive growth, China only has minor deposits of oil and natural gas underground but <u>enormous coal reserves</u>.

To give an idea of China's meteoric growth, here is an example in steelmaking.

	The **U.S.**	Steel production capacity, Tons	**China**
1978	100 Million		20 Million
2020	95 Million		1.2 **Billion**
	Change - 5%		**Change X 60**

Although China consumes huge quantities of mostly imported oil and natural gas, its growth is essentially powered by indigenous coal mining. The following figure synthesizes the past and short term future of coal production and thus burning, highlighting the growing share of China, India and Indonesia since 1990.

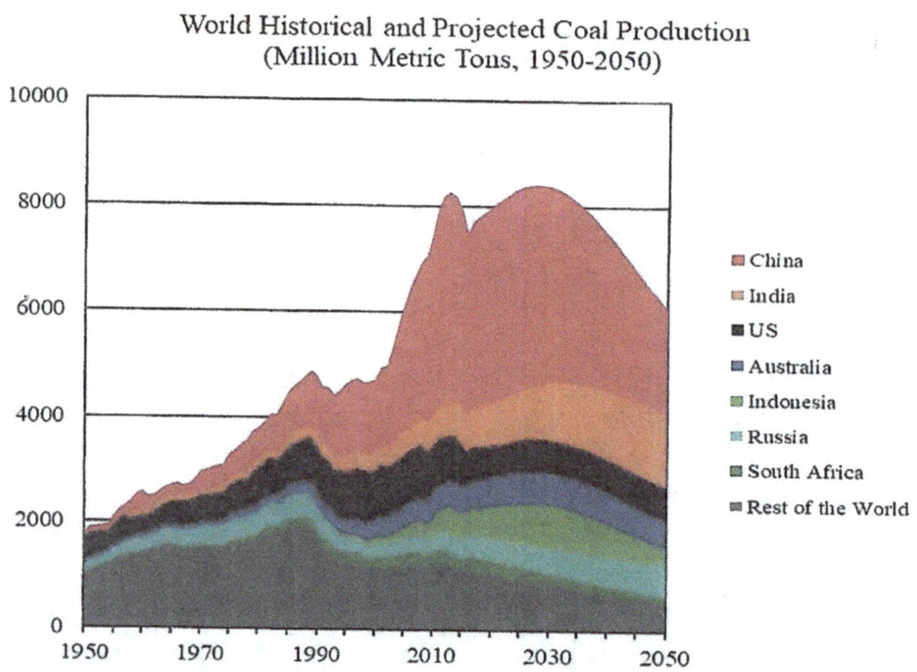

Fig 72. The U.S. coal burning peaked around 2000. China's will peak around 2030. India and other Asian Countries, Indonesia, Vietnam, etc.. will peak later, not what we see in this diagram after 2030.

From an article titled "Coal Fired Power – Brown Elephants"[64] in the Economist of June 30, 2020, we extract a telltale diagram that encapsulates remarkably well the world's increase in coal burning power plants despite the IPCC and reductions in Europe, the U.S. and elsewhere: It is all coming from Asia.

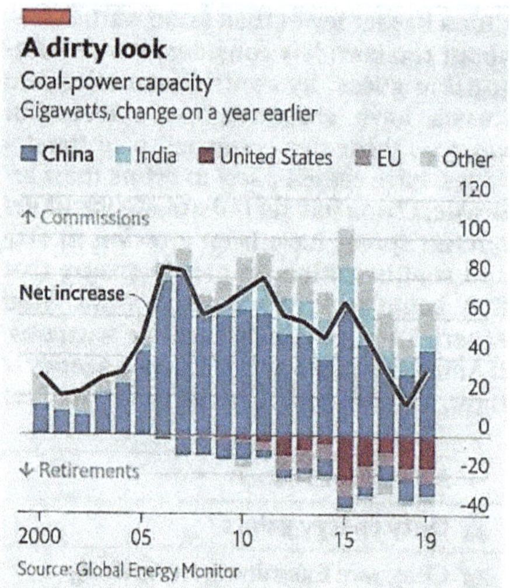

Fig.73. Number of coal burning plants either commissioned, above the red zero line or retired, below the line, every year. Notice the absolute dominance of China, particularly after 2005 and the increasing shutting down in the US and EU. Most important for the future is the growing participation of India and "other" most of these in South East Asia, Vietnam, Indonesia, Malaysia, etc..

In addition to this south east Asian run-away coal burning capacity increase, let us point out the 2011 Nuclear Fukushima disaster. The net result for Japan first, Germany second, was a large return to coal burning power plants as clearly shown in this diagram for Japan.[65]

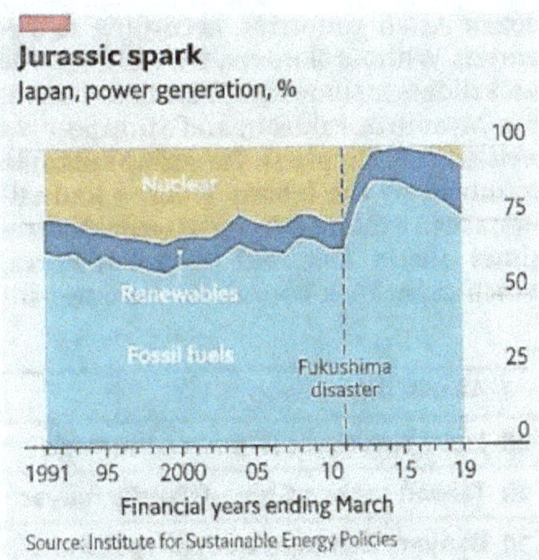

Fig.74. The drastic increase in fossil fuel burning for power generation as a result of the Fukushima nuclear meltdown in 2011. Japan has to import all of its fossil fuels but for power generation, it is almost all from coal.
Source: The Economist, June 13, 2020 [65]

In Germany, Angela Merkel's government decided in 2012 to eliminate its considerable nuclear capacity to avoid a disaster of the Fukushima type. Unfortunately, the only economical replacement available was a thick layer, near the surface, of tertiary lignite "coal" containing only about 15% fixed carbon and lots of impurities except sulfur. So the new lignite burning power plants of Germany have created a new surge in CO_2 emissions. No diagram available.

India, Vietnam, Indonesia, etc..

The next big wave of CO_2 generation after China is from India, Vietnam, Indonesia, Malaysia, Bangladesh, Pakistan, and later, Uzbekistan and the other "stan's". As pointed out on the fig.73 of coal burning plants being commissioned every year, the green (India) and gray (others) are on the rise since around 2010 and will evidently take the lion's share over the next decade or so. The "gray", together as big as China and India in population, represents the countries named above except India. So, in effect, when pointing out to Asia as the current and future CO_2 emitter, we are adding 1.4 Billion (China) to 1.3 Billion (India) to another 1 Billion plus (others) or 3.7-4 Billion humans, over one half of the planet's population of 7.6 Billion and counting.
As a result, over the next 10 to 15 years, coal burning in Asia, for power, steel, concrete, glass, etc.. will rise at about the pace seen in China from 1990 to 2015. This allows us to predict the continuing of the Keeling CO_2 curve to 2035 <u>with no abatement even conceivable</u> and the continuation of the exponential rise.

After 2035, the development of Asia will flatten out which means the final stop to the exponential rise and an INFLECTION in the curve, an "S". This means that, later, the Keeling curve itself will plateau and finally, the CO_2 content will decrease. This more distant future is harder to predict and not important for our purpose.

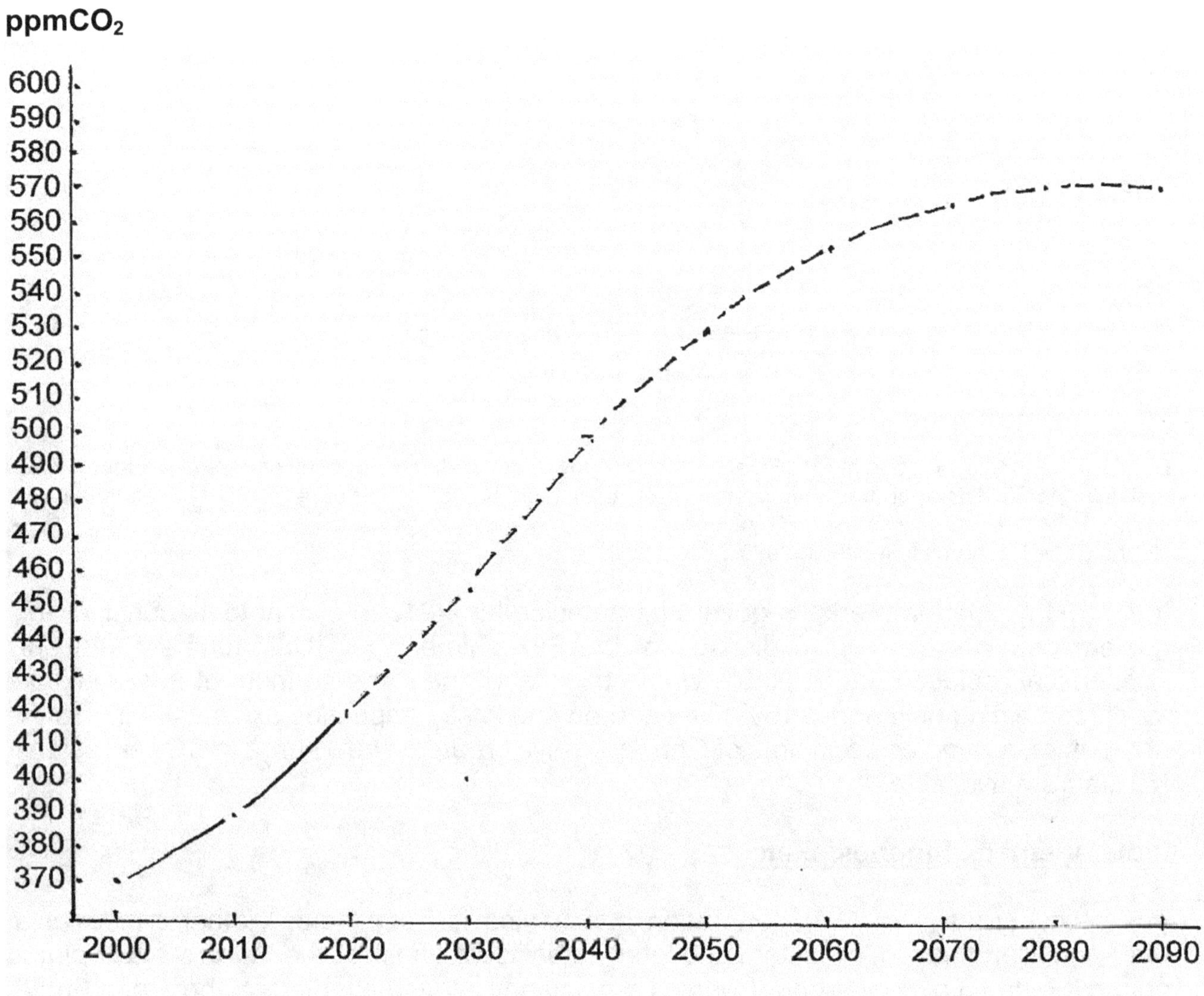

Fig 75. Projecting the Keeling curve beyond 2020 through 2090.

Continuing the exponential slope through 2030 with an inflection point at 475ppm CO_2 around 2035, the curve will top out below 600ppm around 2075 and slowly decrease after that.

Going along with the Greenhouse Gas Effect of CO_2 theory, the preceding pages about the Asian factor demonstrate that , no matter how much progress is made in the rest of the world in CO_2 abatement, **the development of Asia cannot be stopped.** The continuing exponential rise of CO_2 is unavoidable until its inflection point around 2035 at some 475ppm.

The encouraging note however is the trend that follows. A combination of factors such as:

- The end of half the world's population industrial development.
- The ascent of meaningful renewable sources of energy.
- The demise of the Internal Combustion Engine.
- A possible revolution against underground coal mining, even in Asia.
- The precipitous decrease in oil reserves after 2060.
- Breakthrough in power generation such as cold fusion.
- Breakthrough in power storage: new dry batteries.

will probably accelerate the curbing down of the Keeling curve.

We project the topping of that curve around 2075 below 600ppm, then a slow decrease through 2100 and beyond. By then, of course, the greenhousers predict at least a 2^oC rise over today's temperatures and we predict less than 1^oC rise.

Ironically, by 2100, the big industrial shortage could be carbon and hydrocarbons…

Why do we have to spend those trillions of dollars, euros, yens, yuans, etc. on impossible CO_2 abatement?

I hope to have demonstrated the insanity of that foolish goal, even if CO_2 is the cause of our warming.

No matter what Madrid, Paris and Glasgow concoct, the CO_2 content which has risen to 420ppm today, will rise another 180ppm to top around 600ppm in 2075. The existing curve to 2020 clearly shows that all the progress made in Europe, the U.S., Canada, Australia and others is overshadowed by the Asian juggernaut.

* * * * *

<u>Summary of PART THREE.</u>

By tracing the history of the Greenhouse Gas Effect of CO_2 to its well known originators, it appears logical, even inevitable, that the Greens first, then the general public would swallow the fiction hook, line and sinker. The wild assumption of CO_2 being at the core of our warming was presented as fact first by a highly respected Swedish scientist and some sixty years later, also as a fact, by a highly popular science maverick. This, of course, hoodwinked the media who have little or no scientific education. It could only be debunked by older, pre-greenhouse scientists but nobody at the time realized the political and financial damage it would generate later.

The unquestionable facts of Medieval Warmth and mid-millennium Little Ice Ages **negate ipso facto** any possibility that CO_2 would be the warming agent. Indeed, there was no increase in CO_2 1000years ago and no sudden plunge in CO_2 to account for the sudden cooling in 1300.

Realizing around 1991, this massive obstacle to the hoax, someone at the IPCC wrote in an email: "We have got to get rid of Medieval Warmth and Little Ice Ages"! Sure enough, 7 and 8 years later, a team of zealous greenhousers cherry picked proxy data and accomplished the tour de force of erasing any temperature rise in the middle ages and any dip from 1300 to 1700. Jumping from the wild assumptions of the originators, these pseudoscientists crossed the line into deception. There is no law against deceptive science so skeptics have no power to debunk the hoax.

All this Greenhouse talk would be moot if no money was involved. However, as everyone knows, "Climate Change" costs a pretty penny to rich democracies. From zero before 1980, the cost of "Climate Change" rose exponentially. The simple, logical question of the validity of the Greenhouse theory is never discussed in the press or the social media. The skeptics are universally excommunicated as obstacles to the salvation of the planet.

From the onset in 1988 of the Intergovernmental Panel on Climate Change at the United Nations, the IPCC, the total budget for CO_2 abatement soared from millions per year in 1990 to hundreds through 2000, billions by 2010 to trillions of dollars, just for the U.S. in 2020. If the 2030 Green New Deal is politically adopted as is, it would cost just the U.S. from \$33 to 40 Trillions by 2030. This is what climate change damage is to our economy. Disastrous!

From the Keeling curve of CO_2 content in the air, it is unquestionably clear but conveniently ignored by the media, that its exponential shape of growth has not been impacted so far. Neither the trillions on abatement nor all those climate change conferences managed to even begin to slow the curve year after year.

CO_2 keeps rising exponentially and inexorably in our atmosphere. All the IPCC and greenhouse apostle's efforts are totally wasted.

The secret?

Asia and its unstoppable industrial development! So far China emits the lion's share of CO_2 but, until 2035, India, Indonesia, Vietnam, Malaysia, Bangladesh, Myanmar, Pakistan and the other "stans" will take the lead in CO_2 emissions. Then, past 2035, the curve will naturally flatten out. So, even if CO_2 is the cause of warming, there is absolutely nothing that can reduce its exponential growth in our atmosphere, no matter what we try!

<u>Conclusions of PART THREE.</u>

This entire Greenhouse Saga technically boils down to the doubling of CO_2 in our air from 0.03% in 1850 to 0.06% around 2075 while having zero effect on our climate and substantial benefits for the world's vegetation.

As a result: - Stop spending trillions on CO_2 abatement: Totally wasted!

- Start spending trillions on **adaptation** and the **environment!**

- Bring the pseudo-scientists of the IPCC to reckoning.

* * * * * * * * * * * *

PART FOUR

The Reckoning

Between CO$_2$ & Sunspots

By 2030

Overview

Trillions of dollars and Euros could be saved by the Dalton Minimum "scheduled" to bottom out between 2030 and 2040. It has been estimated that this Dalton #2 will cool the planet by up to $1^{\circ}C$ (Part Two, point 11)

Instead of the 2030 Green New Deal expected to cost our rich democracies 15 to 45 Trillion by 2040, we could observe such a climate inversion that the whole greenhouse gas theory may collapse. Indeed, the unstoppable rise of CO_2 should continue the warming.

It is expected that, despite the significant and increasing cooling measured around the globe, the IPCC and most greenhousers will vigorously defend their position with very creative arguments. It might even take us beyond 2030 before the CO_2 fiction finally collapses in the media and minds of this planet.
If no Dalton Minimum occurs, the greenhouse theory will endure, despite the skeptics' firm opinion to the contrary.
The timing of the Dalton Minimum is exquisite because the exponential rise of CO_2 will end around 2035 and the rise itself of the greenhouse gasses will end around 2065.(Part Three, point 11). The warming of our climate, however, should resume strongly by 2055.

1. **<u>The Latest Trend of sunspot Cycles: Dalton # 2.</u>**

After an unprecedented series of very high peaks of sunspot activity from 1935 to 2002, coinciding unfortunately with the rise of CO_2, the following diagram shows a net decline. It started with cycle 23 peaking around 2002.

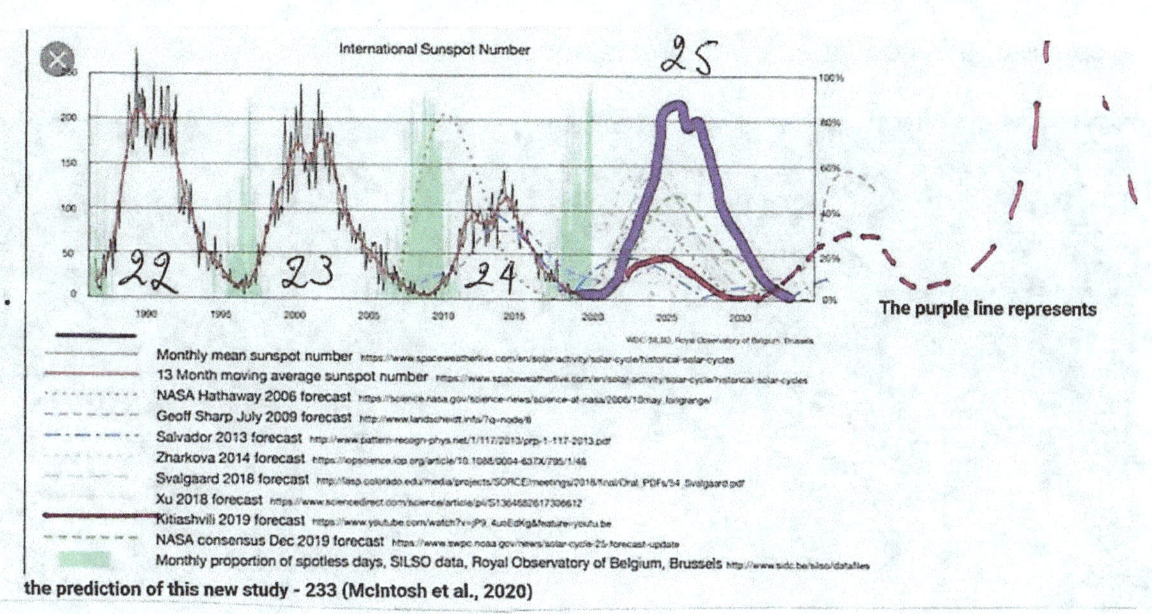

Fig 76. An assortment of predictions for cycle 25 due to peak in 2025. It is clear that the majority of forecasts peg sunspot counts at 50 to 120 in 2025. The McIntosh [66] "purple" line is low, just below 50 and an unidentified heavy blue line peaks higher than 200, highly improbable. The green verticals represent days, weeks, months without any sunspot, introduced by SILSO, the Belgian solar team. I go even lower than the Mc Intosh prediction, 25 to 40 spots at most.

The slowdown in sunspot activity is supposed to result in a global cooling as happened during the first Dalton Minimum.

* * * * *

2. <u>In Late 2020: No Cooling yet: The ALBEDO EFFECT.</u>

From the decreasing sunspot activity of cycles 23 and 24 comes a decrease in magnetic energy hitting our poles. In terms of Watts per m^2, this is significant and should have translated into a beginning of cooling as anticipated by Abdusamatov, Casey, Vahrenholt, etc. from 2006 to 2014. Yet, nothing happened so far. In fact, the warming seems to increase rather than decrease, right through 2020.

The cause of this continuous warming is at our Poles:

In the Arctic Ocean, sea ice reached its minimum extent of 1.44 million square miles (3.74 million square kilometers) on Sept. 15 - the second lowest extent since modern record-keeping began. Credit: NASA's Scientific Visualization Studio

Fig 77. A NASA satellite picture of the North Pole Ocean showing, particularly at the top, rather fuzzy, the largest surface of blue ocean that used to be ice. The date is Sept. 15, 2020.

A Siberian heat wave in spring 2020 started the polar ice melting early with Arctic temperatures 8 to 10°C warmer at the poles than average. The ice surface keeps declining.

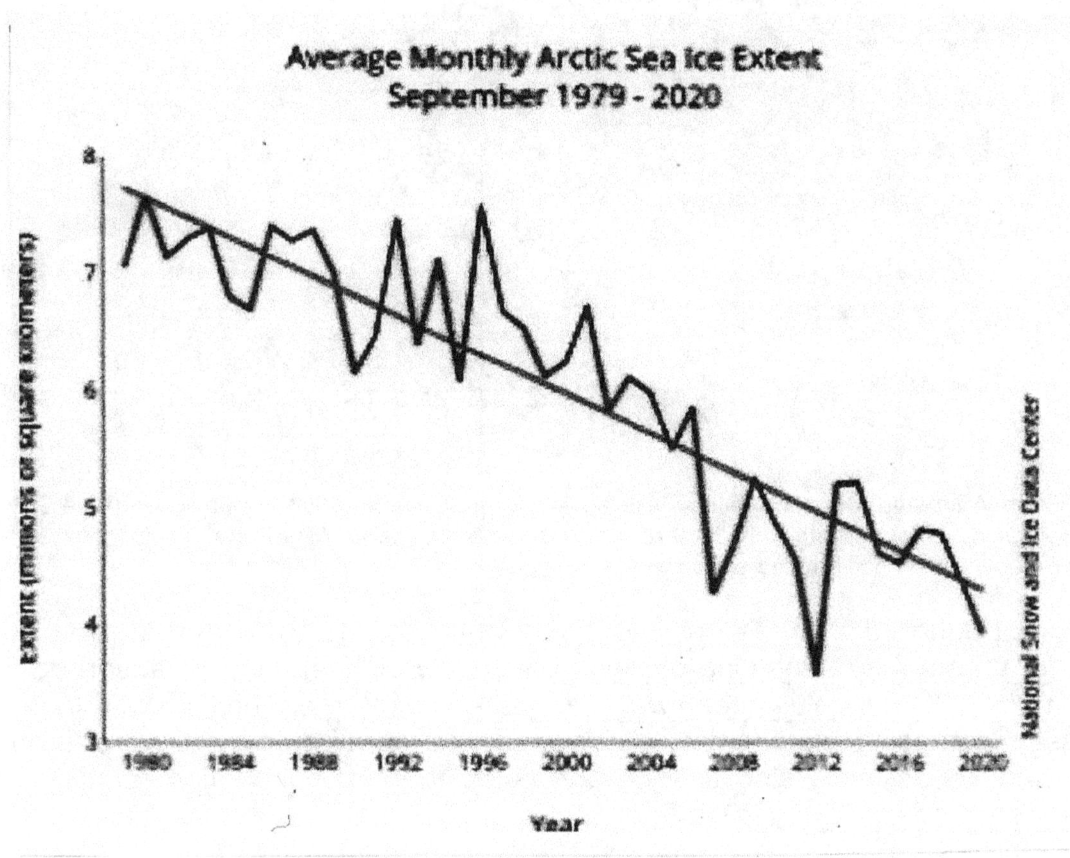

Fig 78. The surface of Arctic Ice has decreased from more than 7 million km^2 in the 1980ies down to less than 4 million in the lowest bottom of Sept 2012 and just above 4 million in Sept 2020. The Arctic Ice surface is down between one third and one half of its previous extent.

When bright white ice is replaced by dark blue ocean an enormous increase occurs in sunlight absorption by the oceans of both poles.

The reflection of sunlight by ice is called the Albedo Effect.

The Webster Dictionary reads: ALBEDO: Whiteness.
Astronomical definition: The reflecting power of a planet or satellite, expressed as ratio of reflected light to the total amount falling on the surface.

The Albedo effect of ice and snow is enormous as clearly shown in the following sketchy diagram.

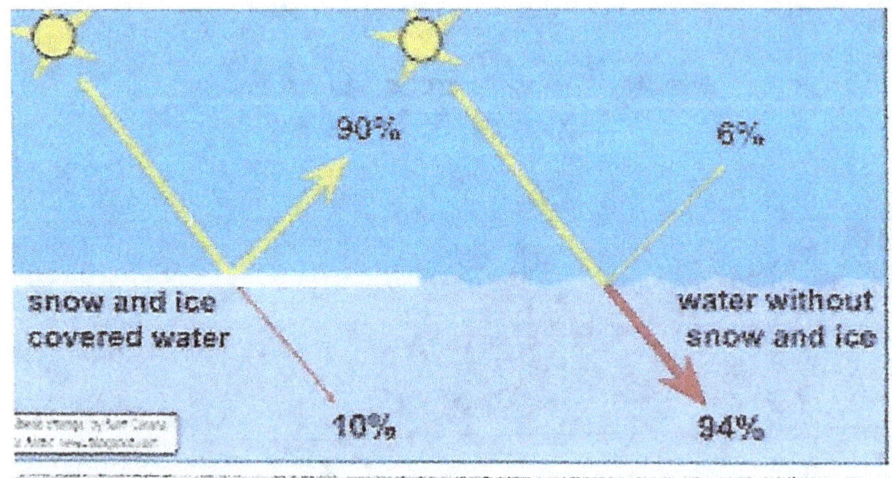

Fig 79.   From Wikipedia this schematic figure shows ice and snow reflecting back to space 90% of the impinging solar radiation, letting only 10 through. Conversely, the dark blue ocean absorbs 94% of the radiation, reflecting only 6% back to space.

So, what happened during the enormous magnetic activity converted to heat over the poles is for a long time a thinning of the ice with only 10% of sunrays absorbed. Then, around 1995-2005, dark blue ocean water appeared over larger and larger surfaces every year, vastly increasing the warming. This time, the warming is by direct absorption of sunlight and keeps increasing to this day, more than compensating for the reduced magnetic energy since about 2010.

* * * * *

3. The Result of the Albedo Effect: Climate Hysteresis.

Hysteresis is used in magnetic systems to signify that the return to zero requires more than reversing the magnetism acquired. It is not just a delay. It means, in climate terms, that the delay in reversing from warming to cooling is caused by the need for a larger drop in magnetic energy than a return to zero.

Following the already weak peak #24 (fig 76), we have experienced a slow descent and months of zero sunspot activity while the warming goes on. It is so unusual that some sunspot climatologists predict a return to the 1645-1700 Maunder Minimum. Indeed, Dalton Minimum #1 from 1795 to 1835 did not feature such long absence of sunspots. We don't go that far but, instead, predict a very low count for cycle #25, maybe as low as 25 spots at the peak.

The preceding paragraph is to illustrate the magnitude of the magnetic energy gap affecting our poles over the last 10 years and more so over the last 2 years.

How long are we in for that hysteresis? Nobody knows! The North Pole's ice melting is such a self-accelerating phenomenon that its reversal may depend on oceanic current cycles suddenly "basculating the Iceberg". All of the solar magnetism climatologists predict that reversal. Many of them, however, predicted that reversal too soon, ignoring or underestimating the albedo effect on our poles.

I predict that the climate reversal will happen for certain before 2030. It is also logical that, the later the reversal, the more abrupt it will be: when the polar ice starts regaining surface, the reduced light energy absorption will enhance the magnetic energy gap and vastly accelerate the cooling.

In effect, it is highly probable that, by 2023, we will feel the beginning of the cooling. Its acceleration over the following 3 years will increase sharply. By 2026, Dalton Minimum #2 will be in full force, making the IPCC and all greenhousers very uncomfortable…

That is what I mean by reckoning!

Every year, we have a simple example of climate hysteresis. Our calendar shows one and a half month delay of our seasons from the sun's position: Summer starts June 21 at Summer solstice instead of being centered on the solstice. If there was no hysteresis, Summer would start May 6, Fall August 6, Winter November 6 and Spring February 6

* * * * *

4. **The South Pole Paradox.**

Superficial climate denying skeptics point at the South Pole as evidence that there is no global warming. "Look at our Pole station, they say, and explain to us why tons of snow keep burying the buildings year after year"! "That is obvious cooling, not warming"!
The snow accumulation is true but strictly limited to a part of the continent roughly centered on the South Pole.

The climate reality at the South Pole is as follows: The Continent of Antarctica with its ice and snow cover is an enormous mass of rock and ice matter stuck at very low subfreezing temperatures. The energy required to melt that ice is far superior to that from the mini-climate warming experienced today. So, in effect, it acts as a moisture sink, precipitating snow from the swirling air current evaporating water from the South Pole's peripheral oceans and spiraling inwards to the South Pole's high plateau.

Conversely, this Little Warm Age does result in giant, flat Rhode Island size icebergs breaking off the continent periphery and melting slowly, away from the shores. This also contributes but more slowly than a the North Pole to the accelerating warming of the albedo effect.

So, to summarize the South Pole Paradox, the warming is evident around its entire periphery but the continent's rock and ice mass acts as a sink for its peripheral water evaporation and keeps accumulating snow on its central plateau.

* * * * *

<u>**Summary of Part FOUR**</u>

The magnetic energy received from the sun and converted to heat at the Poles has been decreasing at an accelerated pace since about 2002. This follows an hyper activity of magnetic sunspots from 1930 to 2000.

This should result in a significant 1°C cooling which as of late 2020, has not started yet. The bottoming out of the cooling should occur between 2030 and 2035, torpedoing once and for all the false greenhouse theory. That upcoming reckoning hangs on the cooling.

The 10 year delay in the beginning of cooling is attributed to the albedo effect of ice melting at the two poles resulting from the hyper warming of the previous 1930-2000 period. This I call climate hysteresis.

The South Pole Paradox is the continuing accumulation of snow on the continent while its periphery is melting.

* * * * * * * * * * * *

<u>**Conclusion of Part FOUR.**</u>

Rich democracies should not wait for the actual reckoning to stop spending trillions of dollars and euros on CO_2 abatement. Instead, they should vastly increase spending on ADAPTING to the next 200 years of warming.

* * * * * * * * * * *

PART FIVE

ADAPT.

Fig 80. "The Need for Climate Adaptation" is clearer than ever"
From "The Economist, May 30, 2020, page 52. Ref [67]

1. <u>CO_2 Abatement Versus Adaptation.</u>

So much focus is concentrated on reducing the CO_2 content of the atmosphere that adaptation to the warming climate receives less attention. It is actually worse: the most active greenhouse apostles are so dedicated to abate CO_2 that they condemn any talk of adjusting to the warming climate as a surrender to the enemy , the cause of that warming.

As a result, budgets for CO_2 abatement are enormous while those for adaptation barely exist:
Estimations for 2017-2018 by Climate Policy Initiative are as follows:

<u>CO_2 Mitigation</u>   <u>Adaptation</u>

$ 537 Billions $ 30 Billions

As concluded in Part THREE, the CO_2 mitigation budget is totally useless even if one believes that CO_2 is the cause of warming. This is because Asia will drive the rise of CO_2 for another 15 years, no matter what we do in the western democracies. So, the budget for mitigation should be 100% transferred to "Adaptation" .
The Adaptation budget should reach $1Trillion per year within 5 years.

*****

To frame the adaptation issue in its historic context, we use the 1,000 year Eddy cycle to super-impose our Little Warm Age, 1750-2250, to the warm Middle Ages, in the fig 81 below.

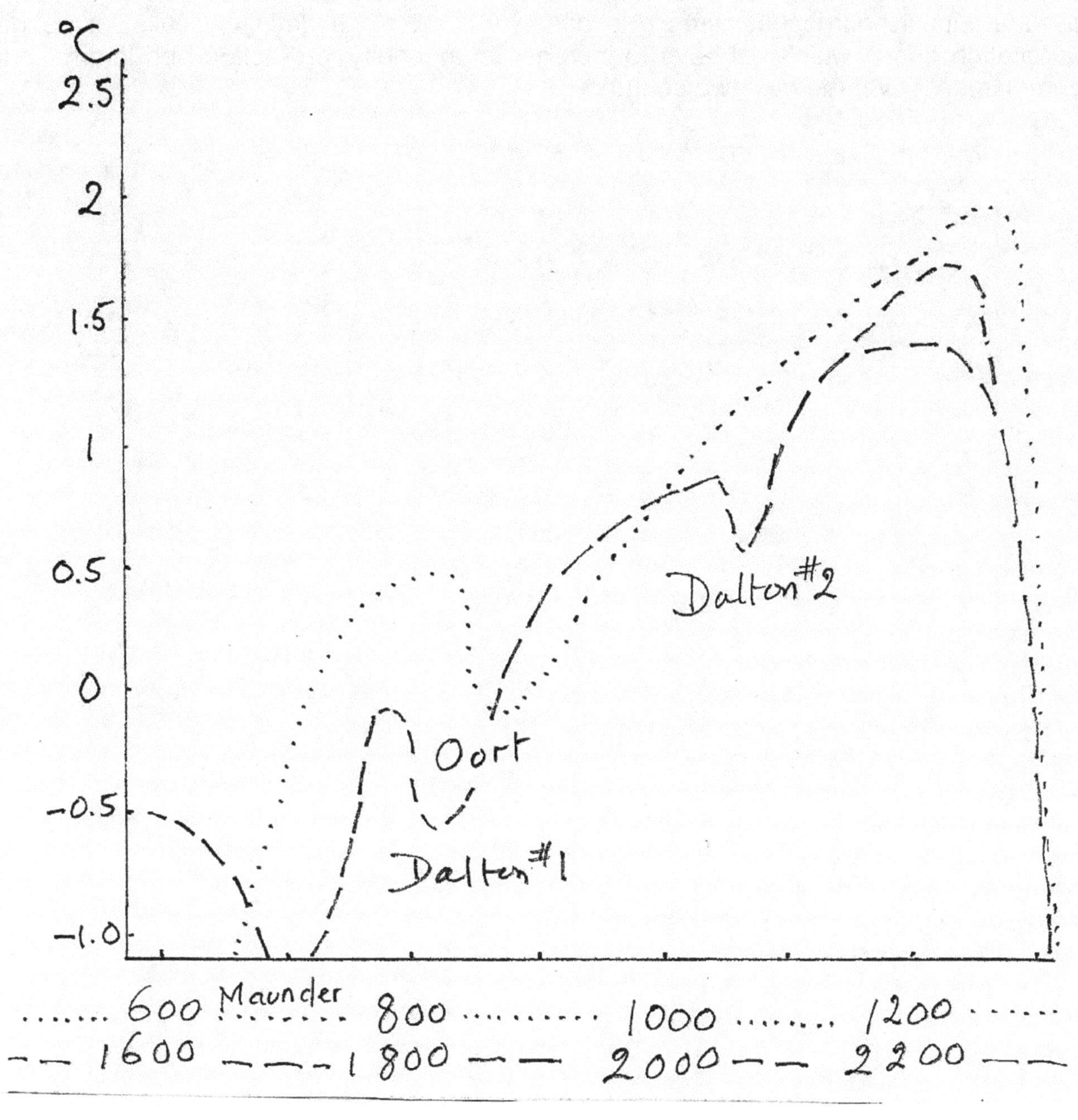

Fig 81. Overlaying our current "Little Warm Age" on top of the Middle Age Temperatures,
we can project the next 230 years.

The preceding diagram shows another 200 years of warming after the second Dalton Cooling, 2025 to 2045. From 2050 to 2250 the additional warming could reach as much as 1.5°C but we predict much less, probably a maximum of 1°C because of the general trend down, cycle after cycle. Minoan summit: 3.3°C; Roman: 3.0°C; Middle Ages: 2.5°C and Little Warm Age: 1.8 to 2.2°C.

So, the further warming we face after 2050 is not great but it comes on top of today's climate with its hurricanes, fires and ocean level rise. It definitely calls for serious adaptation efforts which will have to increase in ingenuity, pro-action and international commitments over the next two centuries.

*****

3. <u>The Dominant Principle of Life on Earth:</u> " <u>ADAPT or PERISH</u>".

Charles Darwin's expression: "Survival of the Fittest" can be translated into adapt or perish. The fittest is, in fact, the one who is preadapted to the changing environment. The changing environment, in our case, is a continuing warming of the climate through the next 230 years.

Refusing to adapt to that change does not always result in immediate disaster. Continuing to build at sea level in Miami, for instance, is not punished by immediate flooding but long term accessibility and thus value of the building will plummet over time.
On the other hand if people persist in living in fire prone forests of California, it may result in early disaster, even death.

Two main modes of adaptation can be distinguished, the second superior to the first:
The first being "Reactive Adaptation": Flooded or burned once but not twice!
Experiencing a climate driven loss, the survival reaction is not to rebuild in the flood zone or the forest, but to take the loss of the land and rebuild elsewhere. Like many of the readers, I am baffled by many stories of intelligent people rebuilding in flood zones along the eastern seashores and around the Mississippi Delta. We don't have to indulge more into personal reactive adaptation: it is self evident.

*****

4. <u>Reactive Adaptation. The South to North MIGRATIONS.</u>

This is unquestionably climate driven. Quite often, the media assign the cause of migrations to political conflicts or to economics. However, the underlaying cause of these massive migrations is the drying up of food supplies which leads to conflicts and economic unsustainability. Nobody had to tell these millions of people to escape famine and survive by moving North. It was simply: Survival of the Fittest!

The two massive occurrences are:
- The Sahel and the Middle East to Europe (30 years).
- Mexico and Central America to the United States (100 years).

Other migrations are either economically or politically driven:
- Workers from India to South Africa under Apartheid
                    to the Persian Gulf States for good jobs
- Exodus from Venezuela for political reasons and starvation
- Exodus from Cuba for the same reasons but decades ago

While alleviating shortness of supplies in the Southern countries left by the migrants, it creates problems in the countries invaded by them. Life style, language, job needs, education and habitat are serious challenges. On the other hand, if reasonably welcomed and assimilated, the migrants refresh and diversify the population of the accepting countries. They also, invariably, generate more children which redresses the often insufficient birth rate of accepting countries.

For the long term, South to North migrations tend to result in a positive outcome for both areas. Only time will tell how good it is.

*****

5. <u>Proactive Adaptation</u> .   **PERSONAL.**

After <u>reactive</u> adaptation, this second mode is far superior because it aims at avoiding any loss by anticipating the danger and acting to circumvent it. For example, I was born in the low lands of Belgium and well aware of the psychological cost of living at or below sea level. Selecting a house location always included that provision: not where even a "thousand year" flood could reach us! That is proactive adaptation. The same instinct against fire danger can be very critical in some areas.

It is difficult to acquire a proactive sense of avoiding climate driven disasters. It requires a conscious effort. One way is to include, in selecting a place to live, a written proviso: No flooding possibility! No fire exposure! No inordinate tornado or hurricane exposure! No possible termite invasion! No excessive exposure to wind gusts, lightning or dust storms. No volcanic or earthquake exposure, etc..

So far we have covered the proactive adaptation issue from the view point of personal living quarters. There is much more.

The more futuristic and advanced proactive adaptation comes next when societies start planning big projects to alleviate the effects of climate change.

*****

6.   Proactive Adaptation: PUBLIC.

In each of our developed countries, at least one University should have a new poly-scientific department called:

"Proactive Adaptation to our Warming Climate"

Focussing young intelligences on this new goal will, with time, produce leapfrogging solutions to problems considered today beyond human control. I suggest as example, greening of the Sahel and of Central America, busting hurricanes before they reach land and, even, abort hurricanes where they originate.. This will be covered in point 9.

Hereunder, we elaborate on some examples of public pro-action.

Rising Ocean Levels.

The Netherlands can and do teach us how to engineer our survival at or below current sea levels. This has been applied to New York City's low laying areas with some success. However, the level will rise further, possibly as much as 10 to 15 feet in the next 200 years. Now, for the U.S., the entire eastern sea coast as well as Florida, Louisiana, and Texas are at stake. It becomes a public necessity to eliminate new private constructions in threatened shore areas. Building permits have been denied, flood insurances rendered illegal in those areas and other publicly mandated precautions like the elimination of drivable access, etc.. All these are negative, punitive steps necessary but insufficient to stop the fashion and craze to build just beyond the ocean waves.

Positive planning should also be considered to crown this effort: The creation of a string of "Shore Parks" adapted in style to the local geography and population.
The first requirement is to push private homes back inland far enough to account for at least 10 feet of ocean rise. The buffer coastal land so made public can be developed into multi-faceted recreation zones. These can include all kinds of commercial concessions such as restaurants and hotels, minicar race tracks, mini zoos, inland lakes for sailing and fishing, beach sailing on wheels, glider airports, etc.. No limits to creative imagination!

Western Fires in the United States.

Despite the fury caused by President Trump's remarks when visiting California during the 2020 summer fires, there was truth in them. Let's face it! The climate keeps warming, drying up large parts of the West and causing a lot more fires in the future.

Adapting to the new normal requires much more than just accusing greenhouse gasses for these calamities.

It is impractical to call for "cleaning up the forests from their underbrush "as being done in Europe since the Middle Ages for home heating. While desirable, underbrush cleaning would be inordinately expensive in most places.

Far more effective, a series of measures should be studied to prevent fire prone areas from being built by private owners. This is not easy. It requires creative ways to convince instead of coerce. It also requires depoliticizing approaches by advocating solid science and good listening to local experience.

In short, future home construction should not occur anywhere near the danger of tree-induced fires. Difficult but doable!

So far, even proactive adaptation as described here above with a few examples is matter of fact science requiring little or no imagination or deep research.

Proactive adaptation has a branch, born less than 20 years ago, which can be called the ultimate future of what man can engineer to improve his habitat on earth:

Geo-engineering!

*****

7. The Ultimate Proactive Adaptation: GEO-ENGINEERING.

Geo-engineering is the new science possibly leading to major charges in local climates. The ultimate goal, of course, is to make conditions more livable in specific zones of habitat.

A key tenet of geo-engineering is: "Think out of the Box"! Any idea, wild at first view, has to be seriously considered before being discarded. And it may have to be reconsidered later if key parameters change.

The ALBEDO effect described in Part IV is a good example of geo-engineering. And, yes, it has already been applied by an international team of researchers around the North Pole. It consists of covering thinning off-white ice with a thin substance with much stronger albedo factor to refract more solar radiation and, thus cool and reinforce the underlaying ice. The goal, however, was not to adapt to the new climate but to mitigate the warming around the poles.

Transfer of Fresh Water to the South West.[1]
This is an obvious example of geo-engineering which is easy to understand and within reach of existing technology. Excess snow melting and rain water of the Upper Mississippi region of the local Red River in North Dakota, Minnesota and Wisconsin to be line piped to the South West.

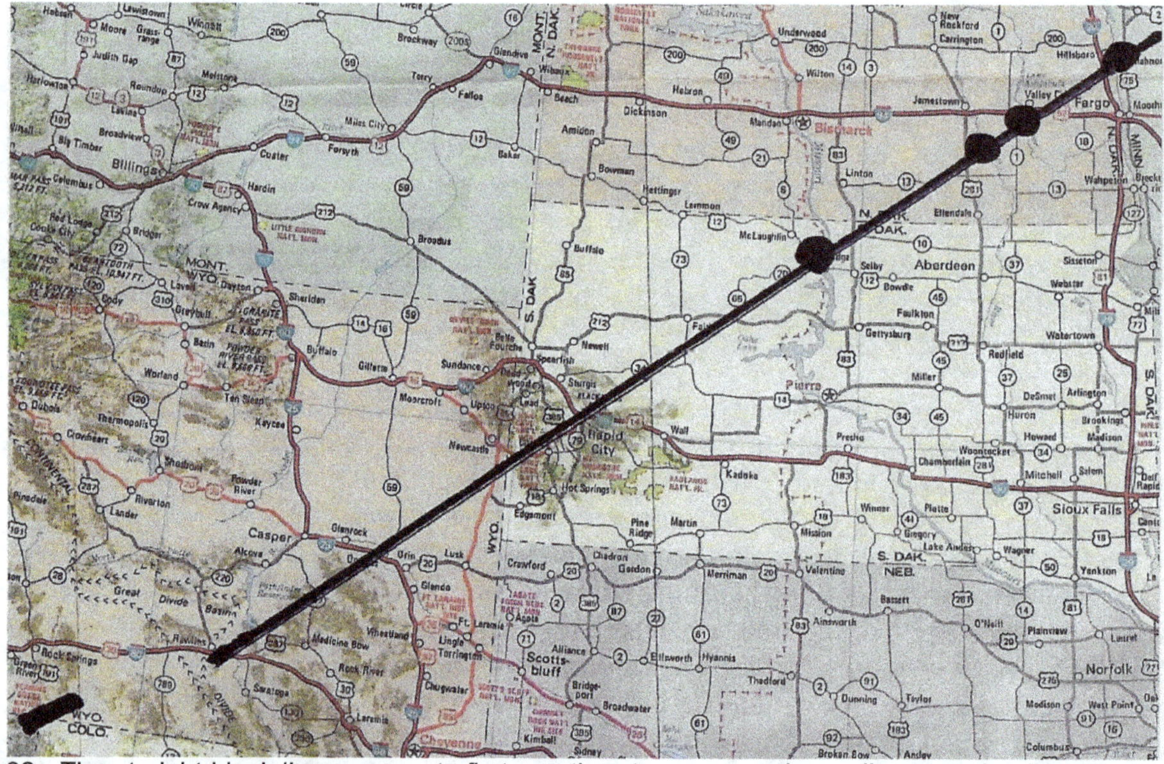

Fig 82. The straight black line represents first one then two or more large diameter pipe lines conveying water from overabundant north central U.S. to parched south west deserts. Parts could be tunneled through the high plateau of Wyoming.

Engineering wise, an important problem will be the altitude at which the water will have to be pumped up to cross the south Wyoming treshhold at some 6,000feet. However, with a tight system, the downhill water gravity, after Wyoming all the way to California, will pull the uphill and reduce the upwards pumping work. Politically, there is heavy lifting to be contemplated.

One approach to avoid an early political burial is to study extensively all the engineering challenges, budget requirements, etc.

The same approach could be suggested to water a good portion of the central Sahel by pumping water out of the Congo River at its further northern point. Here the water pipe lines would be from south to straight north, hundreds of miles. Far more difficult to finance than in the U.S.

I have not found any example of geo-engineering aiming at proactive adaptation but I am offering my own "out of the box" ideas here under.

1/ De-Prime Hurricanes before they hit Land.

Hurricanes are giant pumps of water and air extracting their energy constantly from warm ocean surface waters. The word pump is critical: Pumps have to be primed to start running. In some cases, they can be de-primed, for example, by injecting air in a water pump.

That is the concept! Now comes the hard part.

Two approaches came to mind so far: a) Apply ice cooling on the surface of an approaching hurricane. For example, prepare several retired oil tankers with a full ice load from Arctic or Antarctic sources and have them deliver the floating ice through a newly designed side or backdoor system just ahead of an approaching hurricane.

b) Use explosives dropped inside the hurricane, probably in the eye or immediately around it. The bombs dropped by specially equipped hurricane-resistant aircrafts should be timed to explode above or at the water level. This idea may, at first, be rejected out right but it is a lot cheaper to execute than the first idea. It is also much easier to try out over waters far from land.

2/ About Hurricanes where they originate.

This time the fleet of retired oil tankers loaded with ice from the poles can deploy and drop their load methodically in hurricane generation areas to cool the ocean surface, disabling the original pump before it starts.

3/ Greening the Sahel: Geo-engineering the Gulf Stream.

That same large area of the Atlantic Ocean West of Africa and North of the Equator is the key to the drying up of the Sub-Sahara: the Sahel.

It would be a major accomplishment of geo-engineering to redirect air currents from that ocean area to the East instead of the West. The re-directing may be accomplished at first by the delivery of ice as in 3/ but its systematic and permanent status may require the slowing or the reorientation of the Gulf Stream.

Touching the Gulf Stream would hit an international wall of immediate political opposition, mostly from Europe and Iceland. However reversing the migration from the Sahel to Europe back to the Sahel could make Europeans think harder.

4/ Slowing the Gulf Stream.

Off the coast of Florida where the Gulf Stream is the closest and most narrow, a giant array of passive, power generating stainless steel water propellers would use the entire cross section of the Gulf Stream to generate gigawatts of renewable energy. World politics and investment size are enormous obstacles but the ultimate goal may justify at least a serious consideration and competing modelling studies of the issue.

5/ Redirecting the Gulf Stream.

It may occur from just slowing it down but beyond that my imagination fails me although it may become necessary to green the Sahel.

*****

Conclusion of Part FIVE

Adapting to our unavoidable future of climate warming is a very rich opportunity for younger generations of engineers, scientists, organizers and forward thinkers to contribute unexpected but valuable solutions to growing problems.

* * * * * * * * * * *

PART SIX

The Environment.

Let Us Clean Up Our Act!

A Swiss friend suggested that my book should include this part as important to the entire message. And so I did!

World media confuse and lump up environment with anthropogenic climate change. A few desperate coal miners and coal fired power plant operators do fit that confusion. By enlarge however, that association should not continue to exist. The media also tend to accuse climate change skeptics of denying both the CO_2 causing warming and the need to protect the environment. Nothing is further from the truth: All skeptics advocate for increased respect of the environment and I am one of them.
There is one glaring exception however. When greenhouse proponents affirm that CO_2 is a pollutant, skeptics firmly disagree. The emission of carbon dioxide is profoundly essential to the life cycle of this planet. CO_2 is not a pollutant. It is critical to vegetal life thus intimate part of life on earth.

Pro-active improvement to our environment is an exponentially growing challenge to the twenty first century humanity of the world. Far too much attention and trillions of dollars are spent on CO_2 abatement and too little on the environment.

<u>**1. Retire the Internal Combustion Engine: Stop Nitride Pollution.**</u>

Here we rejoin the climate change advocates but for the true pollution reason: Not CO_2 but CO, Nitrogen Oxides and other poisonous gasses. This aspect has suddenly and unexpectedly been amplified by the corona virus pandemic during the spring of 2020. For a short time over China, the satellite detecting nitrogen compounds relayed a significant drop in these chemicals, due to a plunge in truck and automobile traffic over China's vast territory. Retiring internal combustion can be projected to reduce these cancer-producing pollutants, probably by more than 80%.

Easier said than done! Big car and truck makers around the world are deeply invested in the traditional gasoline and diesel engines. They are painfully aware that the writing is on the wall but reluctant to invest the hundreds of billions required. And yet, that's where at least two thirds of the pollution comes from and that's also where the alternative exists: the more efficient, non-polluting electric motor and the lithium-ion battery pack. Granted, there is still a lot of room left to invent and develop a much lower weight and volume battery pack per kilowatt.hour but the existing technology allows a strong start.

As always, it is a non-carmaker, Tesla, who pioneered the electric car era under the visionary guidance of its creator, Elon Musk.

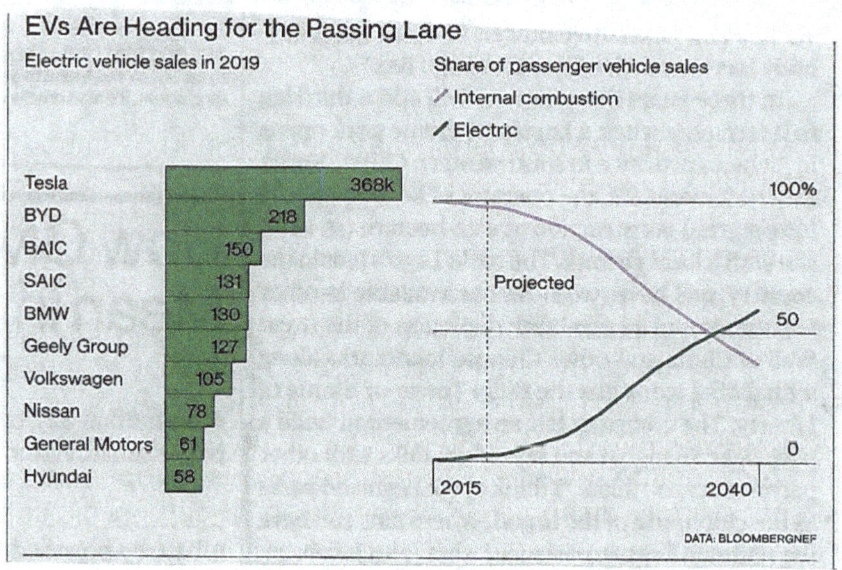

Fig 83. Electric Passenger vehicle sales in 2019 totaling almost 1.5 million units of which about 500,000 were made in China, 470,000 in the US, 235,000 in Germany, 78,000 in Japan and 58,000 in S. Korea. This, of course, doesn't include the many millions of hybrids made everywhere. Source: Bloomberg Business Week.[68] Dec 21, 2020.

Despite many mistakes and years of money losing trial and error approaches, Tesla survived and finally broke through financially since 2019. Its stock on Wall Street clearly projects the future of the electric car by outperforming all other car producers. The above diagram projects a 50/50 replacement of the internal combustion engine by the electric motor around 2035. The rising EV green curve of the diagram seems to underestimate the rise after 2015. The green curve should be much steeper and cross the descending blue one sooner, around 2030.

The truck component is the second highest polluter which adds other pollutants from its large diesel contingent. Its conversion to electric traction seems to be somewhat slower because of the larger energy demand and, thus, battery cost. Midsize trucks made by Rivian for Amazon are a good example of recent moves towards electrification.

In California, Europe, Australia and elsewhere, public and private busses have long leapfrogged the passenger cars with electric motors and battery packs. The conversion curve is far ahead but flatter at this point.

Transoceanic shipping will remain diesel driven for a much longer time, possibly by over 75 years, because ,again, of the enormous energy demand filled by cheap bunker oil.

Airlines also will have a much longer delay replacing the jet propulsion with clean alternatives. Again, the energy volume of jet fuel beautifully stashed away in the wings will take, I estimate, a century or more to be replaced by hydrogen or other alternatives.

In short, by 2075, we should convert close to 100% of cars, trucks and busses to electric power. The retiring of their combustion engines may cut 75 to 80% of its pollutants from our planet's atmosphere. The balance 20-25% remaining, far less tractable to replace, will require new inventions and enormous investments to replace.

*****

2. Rethink Garbage Handling, Collection, Recycling and Burial.

Over the last millennium, the sewer issue took a very long time to resolve and arrive at a relatively acceptable level of control in developed countries. The "third world" however is still beset with open sewers, lack of commodes, etc.. To think that it took, in France, all the way to the reign of Louis XIV to introduce the commode: late 1600's! There is still today, a last step to take with human waste: Chemical conversion.
That was my neighbor's entire focus back in New Castle, PA.

From Google: the world volume of daily garbage generation with almost 7.7 billion consumers is mind boggling. The current world generation is estimated at 0.74 kilo per person per day. That quickly computes to 5.7 million metric tons per day. The generation of garbage is much smaller in poor countries, as low as 0.11kg/day and much higher in developed areas, reaching 4.54kg/day in the US or close to 1.5 million metric tons per day or about 547 million metric tons per year.

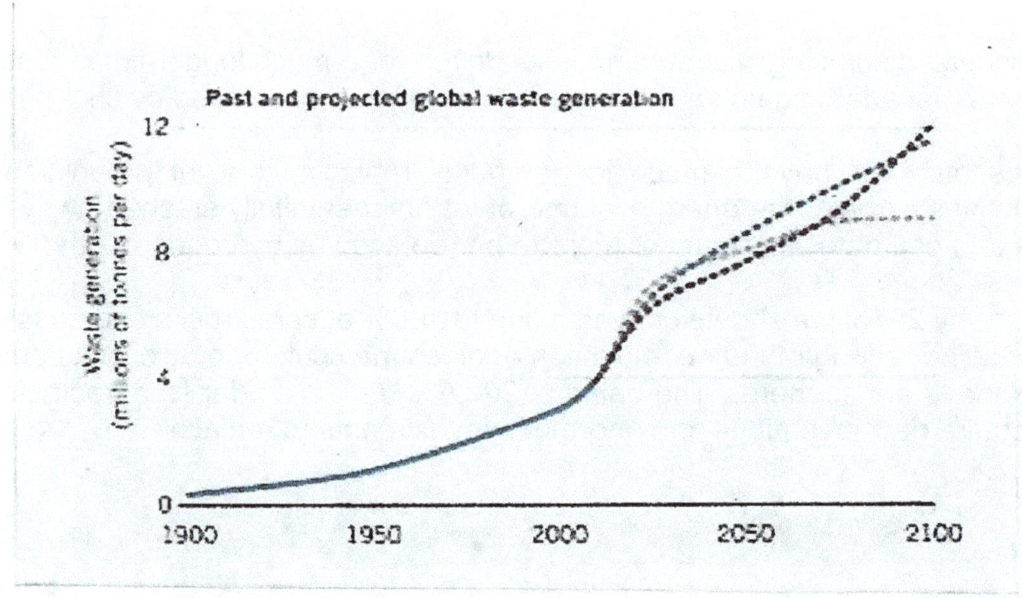

Fig 84.

The worst part of the story is the preceding diagram, on fig 84. Starting from a very low 0.2M Tons/day in 1900 to about 1M Tons in 1950, it climbs to some 3.3M Tons in 2000. Then, the curve steepens sharply and projects about 8M Tons per day in 2050.

This is the challenge we face: Can we rethink the whole issue?
Garbage is not an attractive subject but with a variety of incentives, governments could generate keen interest in some of our best minds.

The rethinking has to go deep and encourage "out of the box" approaches. Even the decades of recycling practices should be re-examined and corrected for better efficiency and results.

Considering the tonnage involved in large US cities for example, here is a politically incorrect, out of the box idea:

Step ONE: Segregate by density the non-recyclable garbage.

Step TWO: - Compress the heavy fraction into dense cubes or spheres.
        - Burn the combustible part of the light fraction and recycle the unburned ashes with the heavy fraction.

Step THREE: Drop cubes or balls into deep ocean bottoms after insulating them from chemical reaction with an appropriate coating to be developed.

I am not suggesting to adopt this idea as is but to consider some of its components as construction blocks for future solutions. It will take highly focused, well educated young minds to come up with the real long term breakthroughs.

Also, of course, the practices developed have to adapt to the local biotope. Low volumes in less developed countries will require lower cost and nimble solutions.

Let us bury our garbage properly before it buries us!

*****

3. Retool Plastics Recycling to Minimize Earth Surface Pollution.

The subject has been made very popular by multiple media coverages of ocean surface collection enterprises with encircling contraptions. Success was limited and required serious adjustments. However the best part of it is that it exposes internationally the magnitude of our earth surface being invaded by non-degradable plastic refuse. Other media focused our attention on streams and open sewers invaded by plastic refuse in cities and suburbs around the world.

The first approach is, emphatically, education! Plastic is over 75% packaging: thin envelopes like shopping bags. It is thus highly compressible and reducible to small volumes, by hand. Uneducated populations tend to just pitch anywhere convenient their plastic refuse like other garbage. Non-biodegradable plastic remains and tends to cover culvert bottoms. Quite a bit of it is dislodged by stormy waters and reaches the ocean where winds and currents tend to accumulate large floating carpets of it at the center of vortices.

That first approach, education, is the hardest part because it requires gentle suggestions on site in thousands of locations. However well inspired media campaigns can be very successful and at low cost.

In developed countries like the US or Europe, plastics do not generally litter the environment but still constitute a headache for recycling. The US could learn from European countries showing future bearing approaches. In Europe, plastic bags are strongly discouraged by charging a fee per bag, for example. People come in grocery stores with their own bags which they reuse many times.

The final disposal of plastics:
Cooling down plastics to various low temperatures can make them as brittle as glass, thus crushable to a fine pulp. That pulp can be burnt at high temperature to reduce the polymerized hydrocarbons to hydrogen and carbon resulting in roughly 50% CO_2 and 50% H_2O by molecules. It can also be used to desulfurize hot metal between blast furnaces and basic oxygen furnaces as explained below.

Successful Annihilation of Plastics and other Hydrocarbons by the Steel Industry.
Back in the late eighties, I read an article by Dr. Peter J. Koros, an earlier colleague at Jones and Laughlin Steel in Pittsburgh, PA., recounting the following story: At J&L's Aliquippa Works, during the night shift, some half-asleep technician had mistakenly connected the hot metal injection lance for desulfurization to the natural gas supply instead of the argon gas piping. Everyone on that graveyard shift was amazed by the suddenly improved sulfur reduction accomplished. After several injection operations someone started checking around. Nothing unusual around the hot metal ladles being injected or at the chem lab. Nothing wrong with the lime magnesium mix used for desulfurization. Finally, the mistaken gas connection was discovered and immediately terminated by the Safety Department. Natural gas is 98% methane, CH_4. When CH_4 hits the 2200°F hot metal at the tip of the lance, it splits instantly into carbon atoms

dissolved in the hot metal and hydrogen molecules reacting as follows with the burnt lime injected and the sulfur impurity:

$$H + CaO + S \rightarrow OH + CaS \quad \text{(in high carbon iron only)}$$

This boosts desulfurization with CaS joining the ladle slag. It never occurred to Dr. Koros that solid polymers could be used but it was obvious to me within minutes of reading the story. Months later, I found the way to put the idea to work. Approaching the President and CEO, Joe R. Jackman, of my previous employer, Remacor, I requested a top level meeting to present the patentable breakthrough. It was in December 1989. The result was patent # 5,021,086 and the recycling, over 5 years, of an estimated 10,000 tons of polyethylene, polypropylene, old tires and an assortment of other used hydrocarbons. Remacor could save 2% magnesium at $1.75/lb replacing it by 1.5% plastic fines at 14 cents/lb. Results, particularly at US Steel in Gary, Indiana, were so competitive that additional patents # 5,873,924; # 5,972,072 and Brazilian #P19809070 were obtained in cooperation with US Steel's Robert Branion to fend off the two fierce competitors, ESM.Inc. and Rossborough.

Unfortunately, after 2-3 years, the suppliers of recycled plastics became very sloppy with the only key specification of minus 40 mesh necessary to avoid injection lance plugging. Neither I or any supplier ever thought of cooling the material way down before crushing it, which would have increased prices to some 25-30 cents/lb but would have saved their business from collapsing. Instead, Remacor's Brian Kinsman knew a good alternative, Wyoming lignite coal fines which completely resolved the difficult injection challenge and retired the plastics for good after 1995.

Twenty five years later, injecting lime, magnesium and flux mixes is still a main stay of desulfurizing hot metal between blast furnaces and basic oxygen furnaces. The distinct possibility still exists to replace part of the magnesium by 100 mesh plastics and old tires first cooled down to brittle condition. This time, recyclers should take the lead, although it is not recycling. It is total annihilation of old hydrocarbons without any secondary polluting effect and with a well established economic incentive. Well handled, that market amounts to millions of tons of used hydrocarbons per year around the world.

Today, plastic is exceedingly cheap because of its origin. A by-product of the petroleum industry, generally starting from ethylene or other root starters. It is then polymerized into polyethylenes, polypropylenes, etc.. By the end of this century, however, most major oil reserves will be exhausted and ipso-facto, render these oil by-products far more expensive. The result will be the disappearance of plastic packaging altogether because other oil by products are far more important than packaging to our modern economy.

*****

4. Replace Fossil Fuels by Renewables and Hydrogen Fusion.

Here also we rejoin the greenhousers but not to avoid CO_2 emissions. In Part Three, we showed that Asia will keep boosting coal burning for at least 10 and, possibly 15 years. Natural gas will keep growing even further in the US and elsewhere because it is so plentiful and flexible, filling the gaps of renewables.

The simple reason that we need to replace oil and natural gas is that their reserves are limited: less than 60 years for oil and less than 100 years for natural gas. Coal reserves are so large that they will be abandoned long before they are exhausted, even in Asia. The hardship of underground mining will eventually, even in China, deter future generations, after 2060, from continuing that hard work.

The complete replacement of fossils by solar and wind renewables is an enormous challenge and, in many countries, impossible. It also requires new breakthroughs in power storage which can amount to terawatts of energy when neither solar nor wind is available.

Fissile nuclear power plants can last another couple of decades and are actively being built in China today. Their cost per kilowatt, however, is so uncompetitive with natural gas and renewables that their long term prospect seems doomed. The life of old nuclear plants in the US, Canada, France and Belgium is being stretched way beyond their original limit and that, of course, saves millions of dollars and euros but increases the risk of a meltdown. Furthermore, when retired, the upfront investment in a new one has become prohibitive. So, unless a new breakthrough occurs, fissile nuclear power seems to be condemned to die out by 2050 or so, except in China.

Hot, Center-of-the-Sun type hydrogen fusion is being developed both in Europe and China with required temperatures and pressures that are mind boggling. I, for one, can hardly believe in the practical cost competitiveness of that approach. Yes, power might be generated but at a breathtaking cost with little prospect of reduction.

Cold, room temperature hydrogen fusion seems totally out of the question. Yet, I for one, strongly believe that it has occurred in Utah and elsewhere about 30 years ago in crystals of palladium. The entire story has been trashed by the most reputable laboratories. [69]

Yet the need for a new source of energy is so enormous and the cold approach so infinitely cheaper than the hot one, that room-temperature fusion power should be reinvestigated thoroughly.

Top notch crystallographers and metallurgists should be an important part of the new teams. Because of the rarity of palladium, other hydrogen absorbing metals should be considered, such as lanthanum.   The LaPd binary diagram has to be reexamined. Lanthanum cost about $1 per lb. Palladium is around $ 2,000 an ounce. La is highly oxidizable while Pd is stable in our atmosphere.

The following book gives an excellent introduction to the subject: "Thirteen things that don't make sense" by Michael Brooks.[69].2008. In its Chapter 4: Cold Fusion, pps 57 to 68.

On November 14, 2019, Jonathan Tennenbaum[70] published in Asia-time, picked up by Google, a must-read article titled: "Cold Fusion: A Potential Game Changer".

In summary, by 2100 we can expect the picture of energy generation to be totally different from today. Fossil down to near zero except gas power plants. Hydropower plus solar and wind maybe 20 to 40%, balance hydrogen fusion of some type.

*****

5. Review the Chemical and Nuclear Pollution Generated by Man.

This last point encompasses the entire damage to the environment caused by our 21st century industry. An important fraction was covered in points **1.** to **4.** But, possibly, more has been left untouched so far in this review, like fertilizers, weedkillers, pharmaceuticals, etc..

It would take a whole book to cover the issue and there are many of them already. The hundreds of waste streams our industry is generating would be drastically reduced if the powerful managers accepted total transparency.

As introduction to the subject, the following book avoids extreme views and offers balance and history of the science:
   "A Climate of Extremes" by Patrick Allitt.[55]. 2015.
   "America in the Age of Environmentalism"
   "Allitt's well written and provocative book has given me more to think about than any other history of the U.S. environmental movement." Adam Rome, author of "The Genius of Earth Day".

****

PART SEVEN

<u>The Long View</u>

Back 1 Million Years
Forth 2 Thousand Years
<u>On Climate</u>

The Past Maxi and Mini Cycles Allow a Fair
Projection of the Next 2 Mini Cycles

1. Building Blocks for Climate Projections.

It is the interweaving of alternances of heat and cold with human history that revealed in Part One the cyclic nature of our climate. The cause of these mini-cycles was extensively developed in Parts Two and Three. The origin of the maxi-cycles was briefly proposed at the end of Part Two. It was concluded that both mini and maxi-cycles are generated by the sun. The fact that the sun is the cause of both cycles in a regular predictable fashion, permits projections with acceptable probability.

The period of mini-cycles is 1000 years; that of maxi-cycles 100,000 years. No wonder humans took such a long time grasping the evidence of these cycles particularly the mini-cycles. Their lifespan is at most one tenth of the mini-cycles and one thousand of the maxis. Only sharp transitions gave humans an occasional chance of reflecting on climate change. 130 years ago air temperatures started to be measured. Barely 33 years ago the current climate craze woke everyone up to the issue. A conscious orientation of mind is required to connect human history with climate.

One of the earliest building blocks connecting hominids with climate was the previous 125,000 years old interglacial, the Eemian, as we will detail later in point 5.

The last sharp rise in heat, 13°C, occurred some 14 to 12,000 years ago and, as we will develop in the Epilogue, triggered the massive enterprise of the large pyramids. It may also have caused some similar enterprises in the Andes of South America.

The epiphany on the cause of mini-cycles came from the last little ice age coinciding with the Maunder Minimum, no sunspots for 55 years. Other building blocks include Ceasar moving his legions to the Scottish border; the greening of the Arabian desert for the Muslims and the 450 years of medieval warming to enable Genghis Khan to ravage central Asia and eastern Europe.

* * * * *

2. The One Million Year Picture: 10 Maxi-Cycles

Backwards to learn forwards.

Thanks to yearly ice accumulation at the South Pole and in Greenland and thanks to the extremely reliable oxygen isotope temperature proxy, we witness 10 maxi-cycles over the last million years (Antarctic only).   As we examined at the end of Part Two, these maxi-cycles show a consistent temperature signature: From the top of the short interglacial there is a 90,000 year asymptotic curve downwards ending at the lowest temperatures, - 7°C to – 10°C. Then a sudden, very steep upsurge to the next interglacial appears on all but the last interglacial: the Holocene.
Starting from primitive, straight walking hominids, our ancestors lived through almost the entire one million glaciation period, far South in Africa. It must have been during the brief preceding interglacial, the Eemian, about 120,000 years ago, that the sudden extreme heat (+13°C) pushed several tribes of Homo Erectus northwards through the "Levant", the near-east, into the cooler Europe of the future. They multiplied and prospered briefly. Then, only a few dozen generations later, the Eemian interglacial gave way to increasing cooling and before they realized what was happening to the climate, they were trapped north of the Mediterranean Sea, unable to go back to Africa. Caught in future Italy, Germany, France, etc.. by the challenging weather, the tough survivors adapted progressively to the worsening conditions with tenacity, ingenuity, intelligence and luck.
Climate was thus the trigger for the game changing transition from homo erectus to homo sapiens.   Surviving long winters in caves encouraged increasingly sophisticated manufacturing of weaponry, everyday home articles and, finally art such as ivory female figures, flutes and broaches to attach animal skin vestments.   In the Lasco caves of southern France, treasures of painted art has, almost miraculously, been preserved depicting hunting scenes, etc.. We also know by now that there were actually two distinct genes of homo sapiens, ours and the Neanderthalensis which was apparently inferior in communication ability. They eventually disappeared around 40,000 years ago.

The coldest part of the last glaciation came at the very end during the period 20 to 14,000 years ago. During this extreme part of the glaciation, already well adapted cave men seem to have accelerated their control of stone cutting and polishing, therefore called the Neolithic era. Also probably woodwork and carving and even possibly, bow and arrow hunting. No trace, of course, of wood and string can be found. Then, within a few decades, the sun must have looked and felt brighter and warmer… It would be fantastic to discover a cave painting relating that event.

* * * * *

3. <u>The Transition to the Holocene. The Dryass. 14 to 12,500 years ago.</u>

The surge in temperatures from the very bottom of -7 to 8°C was fast, a few hundred years. However, unlike the start of all previous interglacials, this one stopped half way , around -4°C for some 1500 years, on a jagged plateau called the Dryass a little flower typical of that semi glacial transition.
Then abruptly again, the Dryass gave way to the full +5°C Holocene within a few hundred years.

This Dryass plateau does not appear in any of the previous transitions from deepest cold to top interglacials. All the steep, almost vertical warmings appear as straight lines in the diagram concluding Part Two. Would that be an indication that something is different in the 10th cycle, announcing also something different for the following interglacial, our Holocene? I believe that, somehow, this Dryass anomaly foretells an abnormal Holocene and it turned out, it did: somewhat cooler, flatter and longer.

* * * * *

4. <u>The Holocene, Our Current Interglacial.</u>

The plus 9°C rise, in a few hundred year from the Dryass to the very top +5°C warmth of our early Holocene generated massive floods on continents worldwide. Several authors connect this accelerated melting of ice sheets with the bible story of Noah and his arch. In the western United States, perhaps the best scientific study of that early Holocene flooding was published in 1993 by Donald K. Grayson[71] in a book named *"The Desert's Past".* One evening with the proper light, I observed from my hotel room in Provo, Utah, a clear horizontal line way up on the steep slope bordering the valley towards the east. Immediately, I inferred that it must have been the shore of an ancient lake and found the book later to learn all about it. The next consequence was perhaps as much as a 60 ft rise of the oceans. Fourteen thousand years ago, not only was England part of the European Continent with the channel dry but the North Sea bottom was dry at least 100 miles North of today's Netherland and today's East England possibly up to York. Except for narrow channels, Norway and Sweden were land connected to Denmark. Dutch fishermen have accidentally dragged human artifacts in their bottom fishnets, miles north of Holland's north shores. These artifacts date back at least 14,000 years when the North sea was a flat land gently sloping northwards.

Before entering further in the Holocene, I suggest for the reader to leap forward and to read the Epilogue. Connecting human prehistory to climate change has its pinnacle in the great pyramids of Gizeh. After the eventful early Holocene, a very long prehistoric period of human evolution is silent. One thing for sure, it takes at least 8,000 years of Holocene for the first awakening of civilizations in China and in the near-east. This 8,000 year gap shows how long it took for mankind to organize its way of thinking and talking in order to learn to live in communities. For the first time also, the interglacial was long enough. All previous interglacials had been 8,000 years or less. If our Holocene had been as short as the Eemian, for example, the plunge into the next deep ice age would have aborted our known civilizations before they could be born.

* * * * *

5. <u>Is the End of the Holocene Predictable?</u>

The length of the Holocene is unquestionably the cause of our prodigious human progress, over its last 4,000 years. However, the reason for that length and for its temperature signature is not clear and unexplained as far as I know. One way to learn about it is to compare the Holocene with its predecessor, the shorter and warmer Eemian interglacial, peaking about 125,000 years ago.

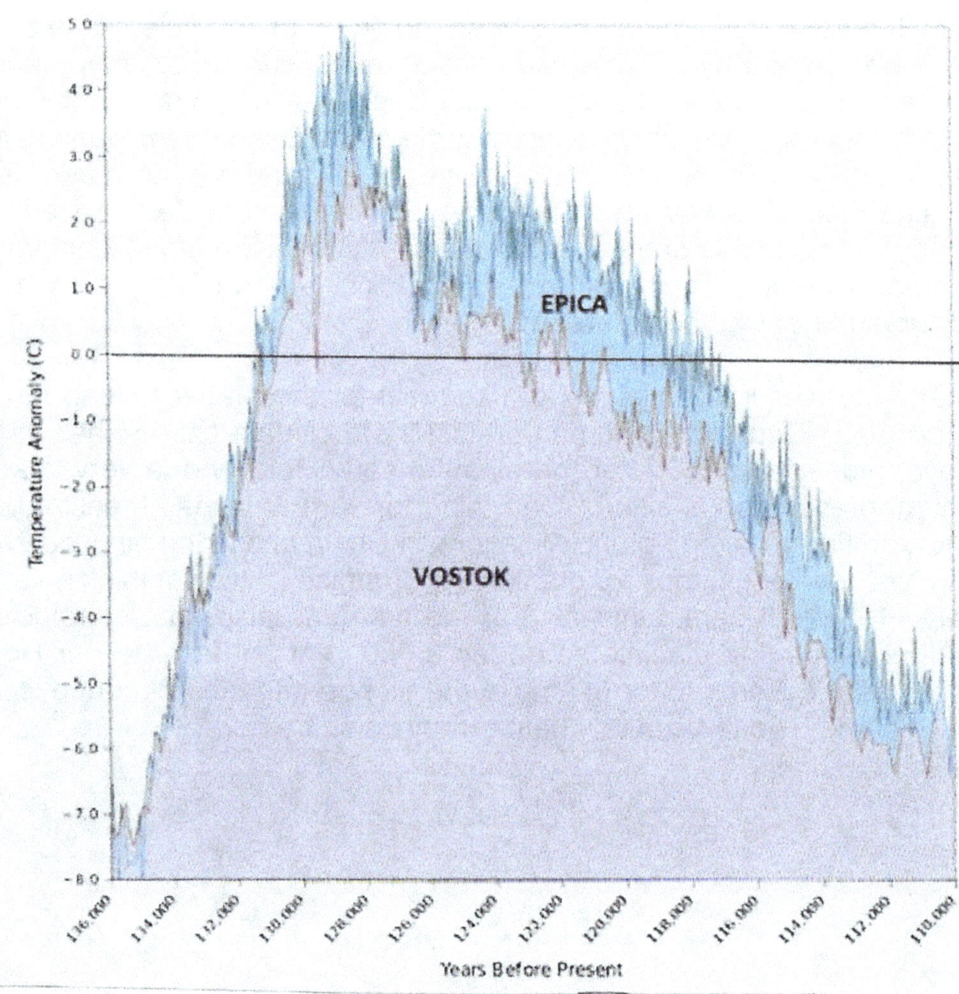

Fig. 85 Both purple and green temperature records of the Eemian Interglacial are from Antarctic ice cores. VOSTOK is a US-Russian-French drilling team and EPICA stands for European Project for Ice Coring in Antarctica. The horizontal line is at 0°C, modified from Dr. Easterbrook.[72]

The zero temperature line in the diagram shows, technically, that the Eemian started 131,000 years ago in both traces and ended 123,000 years ago in the VOSTOK ice cores while lasting another 5,000 years in the EPICA-Greenland ice cores. This means 8,000 to 13,000 years depending on location of the ice cores. So, the Eemian was shorter than our Holocene. Evidence is mounting that the short Eemian was also 2 to 4^0C hotter than today.   Sea levels up to 9 meters-28feet-higher than today, flooding almost all Netherlands and Denmark and half of Belgium today. (More details in Google).

The final message of the Eemian diagram is its dissymmetry. It took from 115,000 years to 125,000 years ago, 10,000 years to rise straight from -8^0C to $+ 5^0$C while the descent took 19,000 years to reach -6^0C. The descent is slower than the rise as we pointed out first at the end of Part TWO. Very important for predictions!

Unfortunately, we can't find the same antarctic diagram for the Holocene as for the Eemian. The following diagram[73] was the closest we can find for a good comparison to the Eemian.

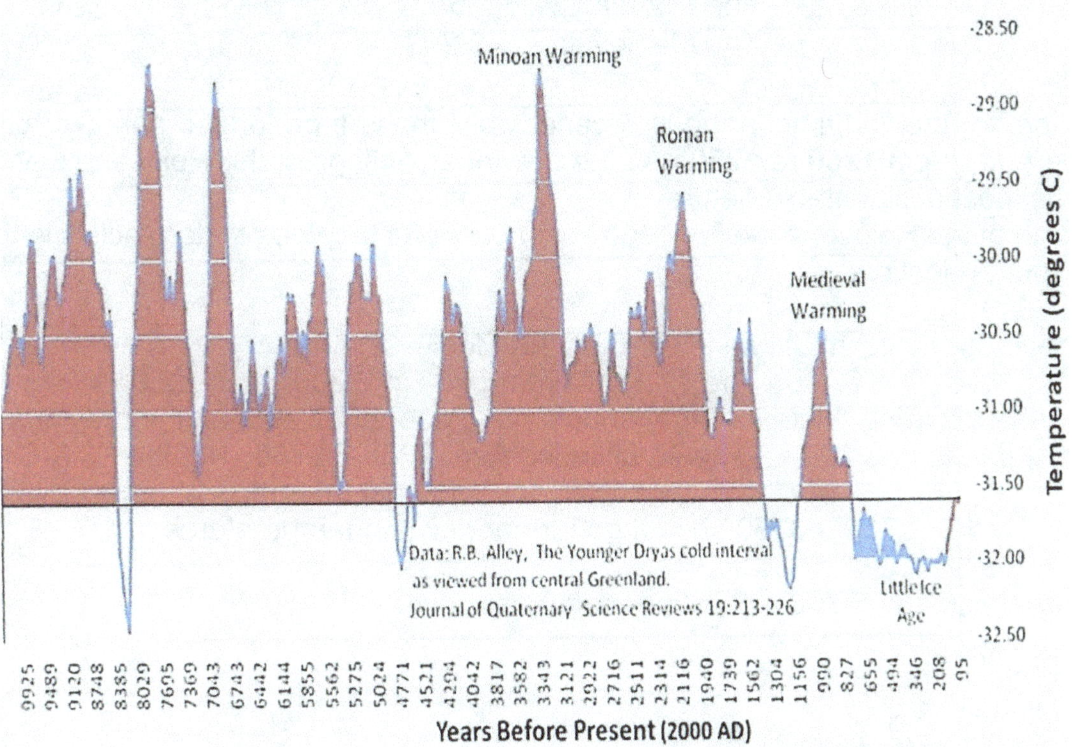

Fig 86. The last 10,000 years of the Holocene (not the complete Holocene!) from Northern Hemisphere Ice cores. [73]. The ordinate temperatures reflect an incomplete translation of the $^{18}0$ to $^{16}0$ ratio temperature proxy. The Holocene is at least 2,500 years older. We miss the all important earlier part of our interglacial to compare with the Eemian.

Comparing the last two diagrams, one for the Eemian and one for part of the Holocene, brings more differences than similarities.

1) The Eemian is like a dissymmetric dome. The Holocene looks more like a flat upper plateau but with sharper ups and downs.
2) The downtrend of the Eemian follows the summit immediately.
    The downtrend of the Holocene only appears over the last 3,000 years.
3) The hot peak of the Eemian, while narrow, reportedly reached 2 to 4^0C higher than any peak in the Holocene, with large consequences.
4) The Eemian resembles all previous interglacials. The Holocene is the one with a different signature, a new trend.

This last point renders any final Holocene predictions less reliable. It indicates a possible new behavior pattern in the sun's radiation mechanism.
The best indicators of the Holocene's demise are the last 3,000 years. It is clear from the last diagram that the 1,000 cycle peaks decrease:

- The Minoan: $+ 3^0$C
- The Roman: $+ 2.5^0$C
- The Medieval: $+ 2.2^0$C

In addition, for the first time in the Holocene, we witness a prolonged period of cold, the Little Ice Age, from 1250 to 1750. All the previous coolings of the Holocene were much shorter, almost accidental in nature.
The length of the Holocene itself, already past previous lengths of interglacials is the final clue for our projections.

First, the fourth peak should be lower than the Medieval:
- The 4th 1,000 year peak in 2,200: 1.8 to 2.0^0C

Second, the following "Little Ice Age" should have deeper dives than the previous one:
- The 4th Little Ice Age, 2250 – 2750: - 2.5 to - 3.5^0C

The Fifth Cycle, 2,750 – 3750 should go further down in temperatures

The Sixth Cycle and 3,750 – 4,750, even further down.

The next glacial age approaches.
    Minoan $+3^0$ C
        Roman $+2.5^0$ C
            Medieval $+2.2^0$ C
                4th   2,200 AD $+1.8^0$ C
                    5th   3,200 AD $+1.0^0$ C
    End of the Holocene: 6th   4,200AD   - 1.5^0 C

Using this descending trend of peaks, it is tantalizingly evident that the Holocene may end in about 2,000 years.
With the Holocene's signature in mind, it is quite possible that the future glacial age may descend further down to -10 or -12^0C at the end.

So, building block by building block, I tried to construct a logic for climate prediction. In short, the result is:

> The next 15 to 25 years: $- 1^0$ C , the Second Dalton Cooling.

> From 2040 to 2200   : $+ 1.8^0$ C, the 4th 1000 year Peak.

> From 2200 to 4200   : $+ 0.5^0$C to $- 4.5^0$C. Holocene ends.

> From 4000 to 100,000 : The eleventh glacial age to $- 12^0$ C.

Harold Ambler was certainly clairvoyant in titling his 2011 book: **"Don't Sell Your Coat"!**

The long view forward is unescapably a progressively harsher deep freeze. However, with new hydrogen fusion energy plentiful by 2200, and barring a giant nuclear or impact catastrophe, humans will be very well equipped to survive the eleventh glaciation.

* * * * * * * * * * *

SUMMARY AND CONCLUSIONS.

Yes, emphatically yes! Our climate is warming up! It is a fact.
Climate-skeptics totally agree. Yet they are often accused of being "climate deniers".
They don't deny the warming. They, we, just disagree on the cause of that warming. The true deniers of the warming fact are a dwindling number of people with ulterior motives.

In effect, climate change is not in question. Its cause is in question and the purpose of this book.

Unlike other climate authors, I emphasize in Part One the central importance of the cyclic nature of climate. This cyclic behavior is interwoven with and revealed by human history of the last 3500 years. This 1000 year mini-climate cycle (+3, -2^0C) is now called the Jack Eddy cycle (Part TWO). In 2020, we were just 20 years past the middle of our fourth Eddy cycle, 1750 – 2250 AD. The Glacier Bay, Alaska evidence strongly disproves the current belief that the warming only started after 1850. In fact, this warming started 100 years before the industrial revolution.

Matching, in Part TWO, these Eddy cycles with sunspot cycles is solidly grounded in the coincidence between the deepest of three Little Ice Ages and the absence of sunspots between 1645 and 1700 AD. Named the Maunder Minimum 300 years later, this disappearance of sunspots generating an abrupt 2^0C drop in world temperatures is the central evidence of the sun being the climate driver.

Our current peak of warming is caused by a paroxysm of sunspot activity between 1930 and 2002. It is being followed by a minor 1^0C cooling between 2020 and 2045 named the second Dalton Minimum. It is caused by weak sunspot energy and numbers, evident today.

How sunspots affect climate was first addressed in Denmark by Dr. Hendrik Svensmark. Based on the observation that sunspot frequency coincides with a decrease in ^{10}Berylium and ^{14}Carbon isotopes, Dr. Svensmark and his team of solar scientists proposed that cloud formations derived from these isotopes explain the warming and cooling. However nobody, so far, looked seriously at the polar lights as a more direct possibility.

Highly visible polar lights are, I propose, the key evidence that sunspots emit invisible matter with high magnetic energy coaxed on our magnetic poles by the earth's magnetic field. The degrading of the magnetic energy into polar lights and infrared radiation warms our planet from the poles. The media have made that evident over the last 40 years. They show massive melting of the Greenland ice sheet, melting of ice near the north pole and calving Connecticut size ice floes around Antarctica.

In Part THREE, the world's belief that CO_2 + CH_4 cause our warming is shown to overestimate their effect by a factor of 200 for CO_2 and 10,000 for CH_4. Although real, the greenhouse effect of these trace gasses cannot produce more than 0.02^0 C by 2100. The history of the greenhouse theory is centered on Carl Sagan's enormous influence on politicians and the media. The United Nations took over the climate theory by founding, in 1988, the Intergovernment Panel on Climate Change, IPCC, with the sole purpose of proving "scientifically" that CO_2 was the sole culprit of global warming. Skeptics fought a disunited and loosing war. The industrial revolution of China and South East Asia will continue the exponential rise of CO_2 until 2035, no matter what may be proposed at Glasgow in 2021.

The Reckoning of Part FOUR explains how, later in this decade, rising CO_2 and falling temperatures will force another look at climate factors.

Part FIVE shows ways to adapt to another 200 years of solar warming.

Part SIX emphasizes the need to protect the environment, not the climate.

Part SEVEN looks at 2000 years forward and suggests the eventual plunge into the eleventh glacial age.

In a nutshell, the **sun** causes all of our climate changes.

* * * * * * * * * * *

EPILOGUE.

The Ultimate, Climate Driven, Tour de Force: The Pyramids.

The second line of the Prologue reads:

"Early Holocene Energizes Man: The Pyramids"

What is this? The large Pyramids, Cheops, Khephren and Mykerinos at Gizeh near Cairo are supposed to have been built around 2500BC. That is all you hear and read today about them. However, in 1995 and 1996, Graham Hancock [74] published two books essentially pushing back the building time of these 3 pyramids by 8,000 years. The second book co-authored by Robert Bauval is titled ***"The Message of the Sphinx"***. It summarizes the results of their joint research proposing the Pyramids were built:

12,500 years ago.

The Sphinx, a sitting lion oriented to the East, dates its building on the constellation LEO breaking the horizon at the vernal equinox, 10,500 years B.C.
The summits of the 3 pyramids pin-pointed on a map, project the exact position of the 3 main stars of the constellation ORION not today or 4,500 years ago but about 13,000 years ago.

I submit that an additional proof of the timing of the pyramids is the extraordinary climate circumstances 12 to 13,000 years ago. From the very deepest cold of the last "maxi" ice age checked at -8°C, temperatures rapidly rose to +5°C with a brief stop midway called the Dryas. The ice sheet accumulated over 100,000 years on the continents melted so fast that it caused major floods, for example in Utah and Nevada and in many other known inter-lands of the world.

The effect, in some 40 to 80 generations of a 13°C surge could be the true original cause of the Pyramids. Not only were our ancestors suddenly liberated from food and lodging worries, which allowed them to live together, it triggered construction dreams, engineering breakthroughs and extraordinary team work.

They had no idea yet of the written word. That came 7 to 8,000 years later but the language used must have evolved and improved rapidly to express all these new concepts.

Yet, it still is a major mystery today how they could manage to build these pyramids. The mass of precisely cut stones has been estimated at 15 million tons. The original cover, still visible at the top of one of them is smooth-cut limestone for beauty. At the base are enormous slabs estimated at up to 50 tons each, cut and aligned with fraction of an inch precision. Contemporary construction engineers would have their hands full pulling this off with their best equipment.

Despite the inference that the Nile was flowing much higher then, floating all that rock cut somewhere else to land practically on site, it was still a gigantic accomplishment.

With nothing in writing and almost nothing left of the tooling, we can only ascertain that the job required enormous effort, team work and, possibly, extraordinary human energy caused by the sudden warming of the early Holocene.

This, I reckon, is the ultimate example to date of climate making man.

* * * * * * * * * * *

REFERENCES.

1. Luyckx L. A. *"Warming? Yes! Manmade? No!"* May 2014. Amazon.

2. Eddy John A. *"The Maunder Minimum".* The Reign of Louis XIV appears to have been a time of real anomaly in the behavior of the sun.
   Science, June 18, 1976, Vol. 192, N° 4245, pages 1189-1202.

3. Encyclopedia Britannica, 1972, book 1, pages 187-192.

4. Meier Christian. *"Caesar"* 1982. Book translated from German by
   David McLintock, 1995.

5. *"The Roman Decline. Empire besieged".* Amsterdam Time Life Books, Inc.
   1988, page 38. ISBN 0705409740.

6. Easterbrook Don J. *"The Solar Magnetic Cause of Climate Changes and Origin of the Ice Age".* 2019, page 45 and others.

7. Encyclopedia Britannica, 1972, Book 15, pages 5-8.

8. Diamond Jared. *"Collapse".* Book 2005. Chapter 8.

9. Guido Peeters. *"La Belgique, une Terre, des Hommes, une Histoire".*
   Readers Digest Sequoia Book, 1980. Chapter 3. Page 79.

10. Encyclopedia Britannica, 1972, Book 14, page 1116.

11. Easterbrook Don J. Previously referred book, page 39.

12. March 20, 2019. *"Small sunspots".* http://www.solarham.net

13. Flamarion Camille. *"L'Astronomie Populaire"* 1891 Book.
   Based on the Meudon telescope near Paris, France.

14. Lemaitre George. *"Un Univers Homogene de Masse Constante et de Rayon croissant, rendant compte de la Vitesse radiale des nebuleuses extragalactiques"*
   1927. Annales de la Societe Scientifique de Bruxelles.

15. Encyclopedia Britannica. 1972. Book 21, page 785.

16. Galilei Galileo. *"Siderius Nuncius."* 1610. Encyclopedia Britannica 1972.
   Book 9, page 1089 and Book 21, page 785.

17. Ambler Harold. *'Don't Sell Your Coat'.* 2011.
    Surprising Truth about Climate Change.

18. Waldmeier. *'The Waldmeier Effect'.* The higher a sunspot peak, the steeper its rate of rise from bottom to top. 1935. Google 6/ 6/ 2021. The diagram used shows, indeed, slow rises for the Dalton peaks and faster rises before and after.

19. Alvestad Jan. *"A Solar Cycle Lost in 1793-1800: Early Sunspot Observations Resolve the Cycle's Mystery."* August 1,2009. Using the newly recovered solar drawings by the 18-19th century observers, Staudacher and Hamilton, we construct the butterfly diagram, the latitudinal distribution of sunspots in the 1790s. The sudden systematic occurrence of spots at high solar latitudes in 1793-96 unambiguously shows that a new cycle started in 1793, which was lost in the traditional Wolf sunspot series. This finally confirms the existence of the lost cycle that was proposed earlier, thus resolving an old mystery.

20. Usoskin Ilya G. Sodankylae Geophysical Observatory (Oulu Unit) University of Oulu. Finland. Publication date: 2009.

21. Vukcevic Milvoje A. *'Equation Evidence of a multi-resonant system within Solar Periodic Activity'.* Rodoslov Bratstva. About 2006.

22. Casey John L.: *'Dark Winter'.* 2014. How the sun is causing a 30 year cold spell.

23. Casey John L.: *'Cold Sun'.* 2011. A dangerous hibernation of the sun has begun.

24. SILSO Graphics. Royal Observatory of Belgium. Nov. 1, 2019. http://sidc.be/silso

25. Vahrenholt Fritz & Lüning Sebastian. *'The Neglected Sun'.* 2014.
    Why the Sun Precludes Climate Catastrophe.

26. Eddy John L. *"The Sun, The Earth and Near Earth Space."*
    A Guide to the Sun-Earth system. 2009. NASA.

27. Singer S. Fred & Avery Dennis T.:*" Unstoppable Global Warming Every 1500 Years".* 2007. New York Times Bestseller.

28. *'The Gaseous Megabet'.* The Economist, April 24, 2021, page 54 near bottom.

29. Hewgill Edith. *"Art Work".* The best rendering found of Carl Sagan's Magician Scientist Power. The Smithsonian, March 2014.

30. Sagan Carl.: *'The Planet VENUS'.*
    SCIENCE, 24 march 1961, Vol. 133, N° 3456, pp 849-856.

31. Pollack James B. and Sagan Carl. *"The Infrared Limb Darkening of VENUS".*
Journal of Geophysical Research, September 15, 1965, N°18, pp 4403-4426.

32. Sagan Carl and Pollack James B.: *"Anisotropic Nonconservative Scattering and the Clouds of VENUS".*
Journal of Geophysical Research, January 15, 1967, Vol. 72, N° 2, pp 469-477.

33. James Lincoln. *"VENUS, the Masked Planet"* 2011. Small fascicule published by Gareth Stevens Publishing. Front Cover.

34. Taylor-Butler Christine. *"Planet VENUS".* 2014. Venus has more than1600 major volcanoes.   Childrens Press.

35. Fone Joe. *"Climate Change : Natural or Manmade?"* 2013

36. Carl Sagan and Mullen George. *"Earth and Mars: Evolution of Atmospheres and Surface Temperatures"*
SCIENCE, New Series, July 7, 1972. Vol 177, N° 4043, pp 52-56.

37. Mann M.E., Bradley R.S., Hughes M.K.: *'Global-scale Temperature Patterns and Climate forcing over the past six Centuries".* NATURE, 1998, 392, pp 779-787.

38. Mann M.E., Bradley R.S., Hughes M.K.: *'Northern Hemisphere Temperatures during the last millennium: inferences, uncertainties and limitations".*
Geophysical Research Letters, 1999, 26, pp 759-62.

39. Montford A.W.: *"The Hockey Stick Illusion. Climate Gate and the Corruption of Science".* 2010. Stacey International. Independent Minds.

40. Mann M.E.: *"The Hockey Stick and the Climate War"* Dispatches from the Front Lines. 2014.

41. Gore Al.: *"Earth in the Balance".* Ecology and the Human Spirit. 1992.

42. Gore Al.: *"Inconvenient Truth".* Movie, 2006 and Book 2008.

43. Mc Intyre S., Mc Kitrick R.: "*Corrections to the Mann et al. Proxy Data Base and Northern Hemisphere average temperature series.*
Energy and Environment. 2003, 14, pp 751-771.

44. Horner Christopher C.: *"The Politically incorrect Guide to Global Warming"* 2007.

45. Sussman Brian.: *"Climate Gate".* A veteran Meteorologist Exposes the Global Warming Scam. 2010

46. Michaels Patrick J. and Balling Jr. Robert C.:*"Climate of Extremes. Global Warming Science They Don't Want You to Know."* 2010.

47. Gerondeau Christian.: *"Climate, The Great Dilusion. A Study of the Climatic, Economic and Political Unrealities."* 2010 .

48. Michael Patrick J. Editor.: *"Climate Coup. Global Warming's Invasion of our government and our Lives."* 2011.

49. Laframboise Donna. *"The Delinquent Teenager who was Mistaken for the World's Top Climate Expert."* The IPCC exposed. 2011.

50. Inhofe James, US Senator. *"The Greatest Hoax: How the Global Warming Conspiracy Threatens Your Future"* 2012.

51. Spencer Roy W.: *"The Great Global Warming Blunder. How Mother Nature Fooled the World's Top Climate Scientists."* 2012.

52. Montford A. W.: *"Hiding the Decline: A History of the Climate Gate Affair."* 2012.

53. Ball Tim, PhD.: *"The deliberate Corruption of Climate Science."* 2014.

54. Steyn Mark Compiling & Editing.: *"A Disgrace to the Profession. The World's Scientists – in their own words- on Michael E. Mann, his Hockey Stick and their Damage to Science."* 2015.

55. Allitt Patrick.: *"A Climate of Crisis, America in the Age of Environmentalism."* Chapter 9 on Climate Change. 2015.

56. Pierce Lawrence E.: *"A New Little Ice Age has Started. How to Survive and Prosper During the Next 50 Difficult Year."* 2015.

57. Wrightstone Gregory.: *"Inconvenient Facts. The Science that Al Gore doesn't want you to know."* 2017.

58. Madden Jack.: *"Inconvenient Facts Proving Global Warming is a Hoax. The Common Sense. Facts for the Basket of Deplorables."* 2017.

59. White Sam.: *"A Cold Welcome. The Little Ice Age and Europe's Encounter with North America."* 2017.

60. Sangster M. J. PhD.: *"The REAL Inconvenient Truth. It's Warming but it is Not CO_2 . The Case for Human-caused Global Warming and Climate Change is based on Deceit, Lies and Manipulation."* 2018.

61. Morano Marc.: *"The Politically Incorrect Guide to Climate Change."* 2018.

62. Bunker Bruce C.: ***"The Mythology of Global Warming, Climate Change, Fiction vs Scientific Facts."*** 2018

63. Smil Vaclar.: ***"Energy Transitions: Global and National Perspective & BP Statistical***

64. The Economist.: ***"Coal Fired Power. Brown Elephants."*** June 30, 2020. Source: Global Energy Monitor.

65. The Economist.: ***"Jurassic Spark."*** Japan's Power Generation. June 13, 2020. Source: Institute for Sustainable Energy Policies.

66. Mc Intosh et al.: ***"New Predictions for Sunspot Cycle # 25."*** 2020.

67. The Economist. :***The Need for Climate Adaptation is Clearer than Ever."*** May 30,2020. p. 52.

68. Bloomberg Business Week. Dec 21, 2020. Data: BLOOMBERGNEF.

69. Brooks Michael.: ***"Thirteen Things that Don't Make Sense."***2008. Book. Chapter 4: Cold Fusion. Nuclear Energy without drama.

70. Tenenbaum Jonathan. ***"Cold Fusion: a Potential Game Changer."*** Asia Time, November 14, 2019.

71. Grayson Donald K. : ***"The Desert's Past."*** Book 1993.

72. Easterbrook Don K. : ref.6 page 108.

73. Alley R. B. : ***"The Younger Dryass Cold Interval is viewed from Central Greenland"*** Journal of Quaternary Science Reviews. 19 / 213-226.

74. Hancock Graham & Bauval Robert: ***"The Message of the Sphinx."***1996

* * * * * * * * * *

AKNOWLEDGMENTS.

This book could not have been written without my wife Claire as down to earth counselor and typist. She was infinitely patient with my rewritings, corrections and multiple additions. Legally blind from wet ARMD, I wrote by hand under a magnifying TV system and Claire logged in everything in the computer.

We owe Janet Johannessen a great deal for her help with computer problems.

Many references came from Google and Wikipedia which we recognize as essential contribution. NASA and the Belgian Uccle-Brussels Observatory Solar Team contributed much.

Part Six, the Environment, was suggested, thankfully, by my Swiss friend Victor Wetterwald.

Finally, the 28 books from skeptic authors and independent minds, acquired and read, were of critical help and information.

The most valuable authors were Fritz Vahrenholt and Sebastian Lüning, Dr. John A. Eddy, Dr. Don J. Easterbrook, Harold Ambler, A.W. Montford, Christian Gerondeau, Tim Ball, John L. Casey, Joe Fone, Graham Hancock and Robert Bauval.

* * * * * * * * * * *

RELEVANT BIOGRAPHY OF LEON A. LUYCKX.

Born in 1933, in Leuven, Belgium, Leon graduated from that City's Catholic University in 1957 as Mining Engineer and in '58 as Geological Engineer. After one year as petrographic teaching assistant to Professor de Bethune and not finding a job as field geologist, Leon joined in late 1959, the Liège, Belgian Steelmaker Espérance-Longdoz, S.A. Immediately applying his petrographic microscope skills to the control of non-metallic inclusions in steel, Luyckx unknowingly became one of its few world's specialists. Late 1964, he was promoted Chief Metallurgist at Allegheny Longdoz, in Genk, B., a joint venture between Esperance, Liege and Ludlum, Brackenridge, PA. There he polished his English and in mid-1967 was hired by Jones&Laughlin Steel in Pittsburgh, PA, where he moved with his wife Claire and five young children. All became US citizens in 1979.

Quickly called "Mr. Inclusions" and later "Mr. Rare Earths". Leon plunged into a career of patentable inventions, first at J&L, 1967-71, then at REMACOR in West Pittsburg, PA., 1973-2001. In 2001 he retired in Ouray, Colorado.

In addition to 25 publications and 12 public presentations at steelmaking conventions, Leon earned and co-earned 18 patents.

These are listed below for easy access by Google:

#1.	5/3/72	3,666,452	# 7.	1/19/82	4,311,523	#13.	6/4/91	5,021,086
#2.	4/20/76	3,951,345	# 8.	8/3/82	4,342,590	#14.	8/19/97	5,658,367
#3.	6/22/76	Re 28,878	# 9.	6/28/83	4,390,498	#15.	2/23/99	5,873,924
#4.	10/4/77	4,052,202	#10.	4/24/84	4,444,590	#16.	6/10/99	P1980.9070
#5.	3/6/79	4,142,887	#11.	5/26/87	4,667,939	#17.	10/26/99	5,972,072
#6.	9/1/81	4,286,984	#12.	7/9/91	1,285,773	#18.	2/26/02	6,350,295

Notes: These are all U.S. patents except #12 Canadian and #16 Brazilian.
The big money makers, for the employers, were #1, #4, #5, #8 and #13.
It was all about clean steel production, low sulfur and low inclusion content.

One would wonder why all these career details are shared in a field totally disconnected from climate? Just to demonstrate how far I engaged my energies to be on the front wave of that particular industrial field.

Since 2004, my entire focus has been on climate, including cruises and trips to Alaska, Iceland, Antarctica and Greenland.

Until 2014, the missing link was the awareness of existing good skeptic references which led to my first essay reinventing the 1000 year cycle completely on my own. Since 2014, the 28 skeptic books listed in Part Three enlightened me but I did not have to change the main thesis of the 2014 essay.

I hope this 16 year effort will bear fruit in reorienting budget decisions, saving trillions of dollars to our rich democracies.

* * * * * * * * * *

Polar Lights: The result of sunspot ejections. The alpha and omega of Climate Change.